BADASS LEGO® GUNS

BADASS LEGO® GUNS

building instructions for five working guns

martin hüdepohl

no starch
press

ISBN-10: 1-59327-284-7
ISBN-13: 978-1-59327-284-5

Publisher: William Pollock
Production Editor: Serena Yang
Developmental Editor: Tyler Ortman
Copyeditor: Kim Wimpsett
Photographer: Rike Gössel
Translators: KittyArne and Birgit Fischer

For information on book distributors or translations, please contact No Starch Press, Inc. directly:
No Starch Press, Inc.
38 Ringold Street, San Francisco, CA 94103
phone: 415.863.9900; fax: 415.863.9950; info@nostarch.com; www.nostarch.com

Library of Congress Cataloging-in-Publication Data

```
Hudepohl, Martin.
 Badass LEGO guns : building instructions for five working guns / by
Martin Hudepohl.
      p. cm.
 ISBN-13: 978-1-59327-284-5
 ISBN-10: 1-59327-284-7
 1.  Firearms--Models. 2.  LEGO toys. 3.  Toy guns.  I. Title.
TS534.5.H83 2011
688.7'25--dc22
                                          2010039988
```

BRIEF CONTENTS

CONTENTS IN DETAIL

WARNING

Adult supervision required. Although these guns shoot only LEGO bricks or rubber bands (not bullets), they can still hurt. Always wear eye protection and never point guns at eyes or other people. For maximum safety when carrying any weapon with a loaded magazine in place, the chamber should be empty, and the slide should be closed. Any gun may fire if dropped or struck. This book is unofficial and is not endorsed or authorized in any way by the LEGO Group.

INTRODUCTION

Major changes in weapons technology mark the milestones of human development, whether the discovery of flint knapping, the secret of smelting copper, or the explosive power of saltpeter. You need to look no further than Genghis Khan's mounted archers or the Roman legionnaire's short sword to see how weapons can change the face of entire continents.

So it's no wonder mankind's first toys were blunted spears, swords, and other "for play" weapons. Even today, when you give a child a stick, it inevitably becomes a cutlass or rifle in his hands. This book aims to ignite your own imagination in much the same way. Of course, you won't be building toy swords, spears, or bows—but instead, working guns inspired by modern weapons and constructed from the most versatile building tool available: the LEGO brick.

I hope you enjoy these designs as much as I enjoyed making them. I provide early prototypes of each gun with plenty of building instructions, illustrations, diagrams, as well as a discussion of general gun design tips and tricks. Each gun has a complete bill of materials, and you can read how to find parts in "How to Get LEGO Parts: BrickLink" page 8. You also learn how to make your own building instructions for your own creations in "LDraw: Build LEGO Models on the Computer" page 12.

Parabella
Parabella is the ideal weapon for covert operations: It's nasty, brutish, and short. Thanks to its slim design, it is outstandingly easy to hide and almost as easy to build. The Parabella is semiautomatic and shoots rubber bands.

Liliputt
Some say a good gun distinguishes itself by intimidating others without firing a single shot. This idea has been realized to my great satisfaction with the Liliputt. This tiny but terrifying semiautomatic pistol packs a serious punch with its nine-brick magazine.

Mini-Thriller
A compact crossbow, the Mini-Thriller is the younger sibling of the larger Thriller Advanced model. At the touch of a button, the Mini-Thriller's bow folds, making its width shrink from thirty-five to five studs. In this state you can hide it inside a briefcase or jacket pocket.

Thriller Advanced
Followers of my videos or readers of my earlier book may be familiar with the original Thriller crossbow. And I now present to you the Thriller Advanced. It is a greatly improved sniper model with extreme stopping power, range, and accuracy. But with its sophisticated mechanisms, it demands the highest standards of its builder.

Warbeast
With its 895 parts and a weight of nearly three pounds, the fully automatic Warbeast is by far the biggest and most badass LEGO gun in this book. With its 30-round magazine and high rate of fire, it also has the greatest fire power.

Magic Moth
The Magic Moth is a primitive butterfly knife. Of course, it doesn't have a real blade. Despite that obvious limitation, it's my favorite model: There is no greater fun than brandishing this gadget and learning tricks with it. Originally, the Magic Moth was top-secret. At the very last minute my publisher convinced me to release the construction plans and added them to the book.

NOTE: *Please be careful while assembling and using these models. All of the models in this book really shoot—and really hurt, too. Never point the gun at a human or other living thing. These models are **not** suitable for children.*

MY LEGO DESIGN PRINCIPLES

There are two ways to join LEGO Technic parts with each other: with studs and with pins. Stud connections are stiff and unstable. Pin connections are heavy-duty and elastic, but flexibility isn't always desirable. If you use stud connections and reinforce them with pin connections, you get an assembly that is stiff, stable, and sturdy.

4×1 System brick introduced in 1967, 1.64 grams

4×1 Technic brick introduced in 1977, 1.46 grams

Technic liftarm introduced in 2000, 0.7 grams

Think about which base elements you want to use: the old Technic bricks or the new Technic liftarms. Technic bricks are classic LEGO System bricks with additional holes for sideways connections by way of pins. Technic liftarms, which connect only with pins, are the newest advancement. Lacking studs, liftarms are obviously considerably smaller than bricks, which make them somewhat more versatile, allowing for more compact and complex models.

The advantage of liftarms becomes clear when you compare the two crossbow pistols. Although they have the same functionality, the Mini-Thriller is far more sophisticated than the Thriller Advanced.

Liftarms are particularly suitable for LEGO guns because rubber bands don't tend to catch on their rounded edges, as they do on Technic bricks. But Technic bricks have their advantages, too. First, they are twice as sturdy as liftarms. And thanks to their square shape, they produce a better look, because, they connect to one another without leaving gaps.

The most important requirement for building outstanding models is to have a large selection of parts. As a minimum requirement, you should have available the parts from the table shown on pages 10–11. I own most parts only in black and buy additional colors once the model is nearly finished. Use colors carefully, because a random mix of colors hardly ever looks good. Other than black and gray, I use only one or two additional colors per model.

The optimal color scheme can be found only through trial and error. Trial and error is difficult to perform on a real model because for each color change, at least one part has to be taken out and a different colored part put back in. Therefore, I recommend developing the final color scheme on the computer using LDraw.

NOTE: *In MLCad, you can easily color your LDraw model at will. But watch out: MLCad's large palette doesn't exist in reality. Most LEGO parts are available in only a limited selection of colors, so check the LDraw-program BrickStore or BrickLink to see whether the parts are available in your desired hues.*

One single, large part is preferable to two smaller ones that are joined. Every unnecessary connection diminishes stability.

The best Technic models are the ones that offer high functionality in a small package.

LEGO bricks don't always have to be used with the studs on top. Put the bricks at whatever angle is most sensible for your construction. The Warbeast was built with the studs pointing down because that was the only way to attach the battery case upside down. In the Thriller Advanced, I positioned the bricks so they wouldn't hinder the launching rubber bands in the launch rail.

Although rubber bands may be indispensable for LEGO guns, use them sparingly because they constantly lose elasticity and have to be changed frequently. Wherever possible, use springs or other elastic parts. For example, in the Mini-Thriller (see page 73), I avoid using rubber bands as much as possible.

If you install a motor, try to use a very small gearbox or none at all. Each additional gearwheel takes up engine power. If necessary, use oil to lubricate a stubborn gearbox.

The MINDSTORMS motors have an advantage over the older Technic motors because of their low-loss built-in transmission.

The new, wide LEGO gearwheels (12 or 20 cogs) are preferable to the old ones (8 or 24 cogs) as they are more versatile and robust.

Old gearwheels, introduced in 1977

New gearwheels, introduced in 1999, which I recommend

TOOLS

LEGO allows you to build sophisticated models without tools. Nonetheless, sometimes you may run into problems that call for additional measures. Household rubber bands, without which none of the LEGO guns in this book would work, are one example.

I recommend the following tools:

A **dull knife** to separate sticking LEGO plates.

Needle-nose pliers, **large tweezers**, or a **hemostat** are well suited for tasks such as removing Technic pins or connecting parts in hard-to-reach places. (See page 212.)

Because the LEGO Group releases new parts in nearly every set, we hobby-designers should have no scruples about using **glue** and **sandpaper** to manufacture our own parts. However, I don't recommend gluing together entire models. LEGO parts are much too valuable for that.

Sadly, the LEGO Group hasn't released any bearings, which is why **oil** is an invaluable aid for gearbox constructions and the like. For lubrication, use clear, odorless sewing machine oil.

Removing a Technic pin with a hemostat

Sandpaper

KNOW YOUR LEGO

To build perfect LEGO models, you should know every way to connect any LEGO piece with any other. And in order to know that, you have to know the exact measurement of the LEGO parts. For example, five stacked plates (5 × 3.2 mm = 16 mm) are the same height as two liftarms placed on top of each other (2 × 8 mm = 16 mm).

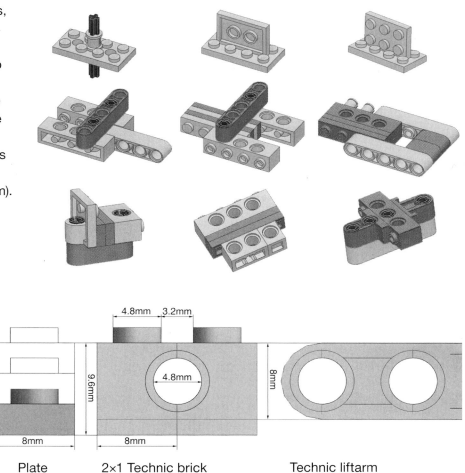

Plate   2×1 Technic brick   Technic liftarm

A lot of cool designs are possible with Technic axle angle connectors. Try it out!

Different ways to create a right angle

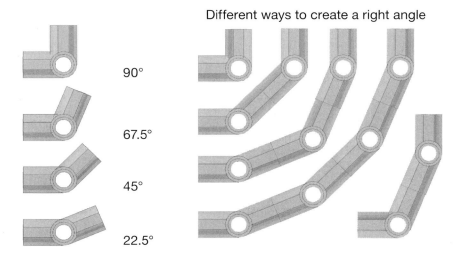

90°

67.5°

45°

22.5°

CROSS BRACES WITH LEGO

Before we had LEGO Technic axles and pins, LEGO parts were just stuck on top of each other with the studs pointing up. Axles and pins made it possible to build diagonally for the first time.

The diagrams below and on the right show holes to connect LEGO Technic liftarms or bricks diagonally. These graphics were useful in the design of the angled magazine of the Warbeast, for example.

The angles 36.87° and 53.13° are especially important. Every five holes, you'll find a common hole on the hypotenuse (the diagonal part).

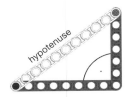

Figure 1: Cross braces with Technic liftarms

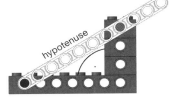

Figure 2: Cross braces with Technic bricks

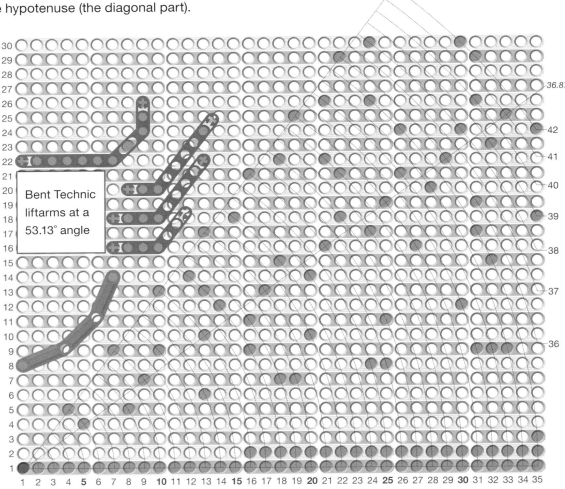

Bent Technic liftarms at a 53.13° angle

Cross braces with Technic liftarms: The red dots signify the possible pin connections for cross braces with liftarms (as shown in Figure 1).

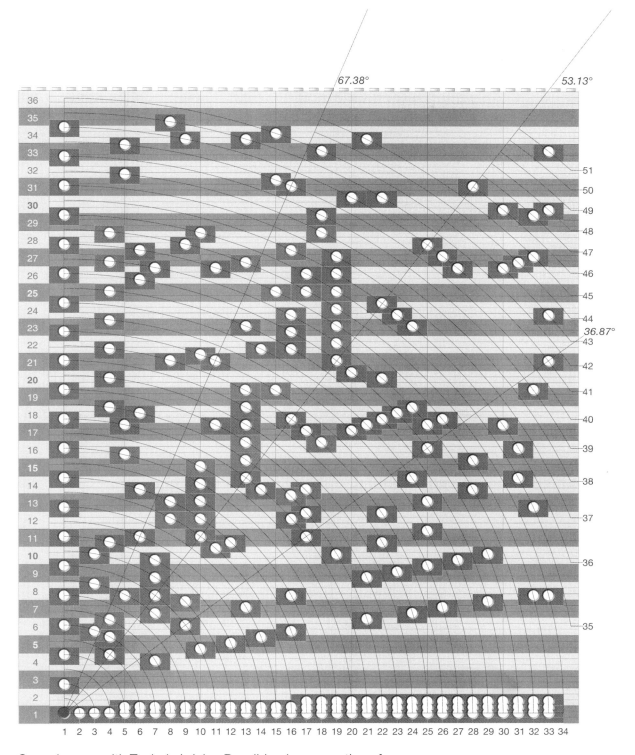

Cross braces with Technic bricks: Possible pin connections for cross
braces with Technic bricks (as shown in Figure 2).

HOW TO GET LEGO PARTS: BRICKLINK

As the popularity of LEGO continues to grow around the world, we can assume that soon the coolest creations will come from aficionados worldwide, not from the LEGO Group. To re-create such fan-created models, you need (1) construction plans and (2) the necessary parts. For the past few years, everyone has been able to competently accomplish the former using LDraw's digital parts library. You can get the latter from BrickLink (*http://www.bricklink.com/*).

In an ideal world, you could just click any LDraw model on the Internet, and all building parts and instructions would just be sent to you—sort of like a specialized LEGO set. Sadly, the necessary bridge between LDraw and BrickLink does not yet exist. Currently, you still have to painstakingly select the necessary parts at BrickLink. The following should assist you with this complex task.

At BrickLink, you can buy any LEGO part or set produced to date; in fact, 80,000,000 parts by 3,500 sellers are currently listed. Unfortunately, the BrickLink user interface is still somewhat old-fashioned and clunky, and buying parts is made difficult by the huge number of sellers, every one of whom has their own pricing strategy and trade terms. Some sellers offer parts at very high prices and others extremely cheap, while other sellers stipulate a minimum order. Occasionally, a seller overcharges for shipping or doesn't ship to every country.

Which Shop Should I Purchase From?

Generally, the more time you spend hunting for the right sellers, the cheaper your order will become. When you need many different LEGO parts, it is advisable to buy from one of the larger dealers, who have assortments of more than 100,000 parts. This saves time and shipping costs. But, if you need just one or two types of parts, possibly in large quantities (21 16×1 Technic bricks in black for the Warbeast, for example), you should buy from the lowest-priced seller. The BrickLink search engine allows you to sort by price.

Exchange Rate

BrickLink prices are always listed in US dollars. Because of the dollar's current weakness, many international sellers make up their own exchange rates that favor the US dollar. International sellers may bill in a currency other than dollars. For example, one might be billed in euros at a 15 percent price increase. For this reason, it can be cheaper to order from sellers in the United States.

Wanted Lists

On BrickLink, you can make a "wanted list." When you do, the system will find the cheapest seller with the best price for you. BrickLink even notifies you when formerly unavailable items from your wanted list are offered, similar to eBay's automated notifier. It sounds great, but as of this writing, the "wanted list" feature is too poorly conceived and takes up more time than it saves. Ideally this will change someday.

Part Numbers

Every LEGO part has a unique number. For example, the well-known 4×2 brick is 3001. The part number depends on shape, not color. These part numbers are used by BrickLink and LDraw, so you can use the BrickLink search and the LDraw database interchangeably. All parts used in this book are listed, with numbers, in the table on page 10. To find a part, simply search BrickLink for the part number and color.

LEGO part 3001

Colors

Some time around the turn of the millennium, the LEGO grays were adjusted. LEGO felt the slightly yellowish tones "dark gray" and "light gray" didn't look contemporary and slowly phased them out, replacing them with the colors "light bluish gray" and "dark bluish gray." It doesn't matter which gray you use for your model, but it looks classier not to mix the warm and cold tones.

Pick A Brick

At *http://shop.lego.com/pab/*, you'll find the LEGO Group's official parts store, Pick A Brick, where you can sometimes find very new or rare parts. Nonetheless, I don't recommend Pick A Brick highly because there is no proper search function, the selection is small, and the prices are inflated to around four times their trading value on BrickLink. However, the Pick A Brick parts are always new.

BrickLink categories

Brick,
Brick, round
Electric, Battery Box
Electric, Motor
Electric, Wire & Connector
Panel
Plate
Plate, round
Slope
Slope, Curved
Slope, Inverted
Technic
Technic, Axle
Technic, Brick
Technic, Connector
Technic, Gear
Technic, Liftarm
Technic, Pin
Technic, Plate
Technic, Shock Absorber
Tile
Wheel

BrickLink offers 20,000 part types in 185 categories. Only the 22 categories listed here are important for this book.

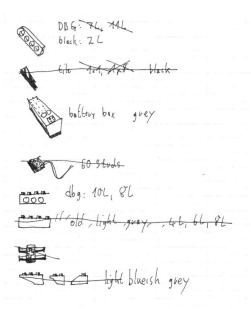

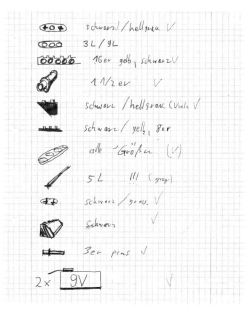

Shopping list: Before ordering from BrickLink, I always take stock of my LEGO creation and draw the parts of which I'm running short. Thanks to the table on the next two pages, you don't have to make drawings; just note part numbers accordingly.

BRICKLINK PART NUMBERS LIST

This parts list helps you find the right LEGO components at BrickLink. Just search there for the part numbers, and remember that the part numbers specify only the part's shape, not its color. For example, if you're looking for a red 1×2 Technic brick, the correct search term is "red 3700." You'll only get results when the order of search terms is first color and then part number.

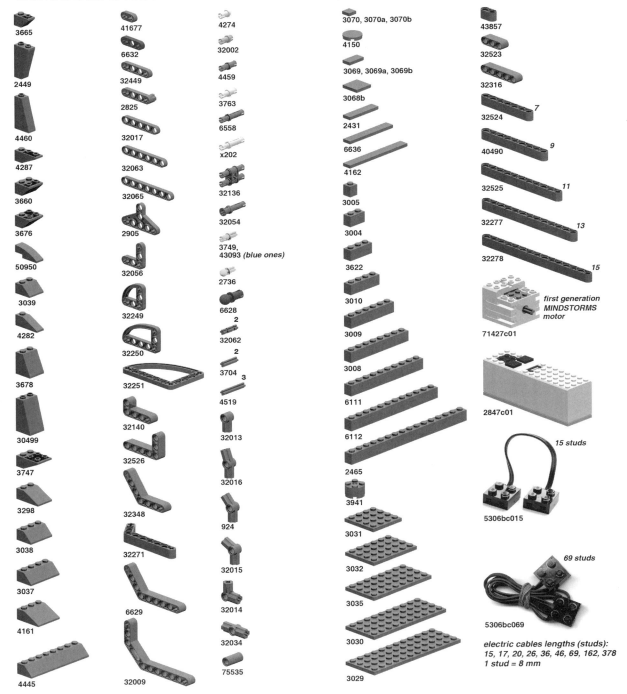

3665
2449
4460
4287
3660
3676
50950
3039
4282
3678
30499
3747
3298
3038
3037
4161
4445

41677
6632
32449
2825
32017
32063
32065
2905
32056
32249
32250
32251
32140
32526
32348
32271
6629
32009

4274
32002
4459
3763
6558
x202
32136
32054
3749,
43093 *(blue ones)*
2736
6628
2
32062
2
3704
3
4519
32013
32016
924
32015
32014
32034
75535

3070, 3070a, 3070b
4150
3069, 3069a, 3069b
3068b
2431
6636
4162
3005
3004
3622
3010
3009
3008
6111
6112
2465
3941
3031
3032
3035
3030
3029

43857
32523
32316
32524 *7*
40490 *9*
32525 *11*
32277 *13*
32278 *15*

first generation MINDSTORMS motor
71427c01

2847c01

15 studs
5306bc015

69 studs
5306bc069

*electric cables lengths (studs):
15, 17, 20, 26, 36, 46, 69, 162, 378
1 stud = 8 mm*

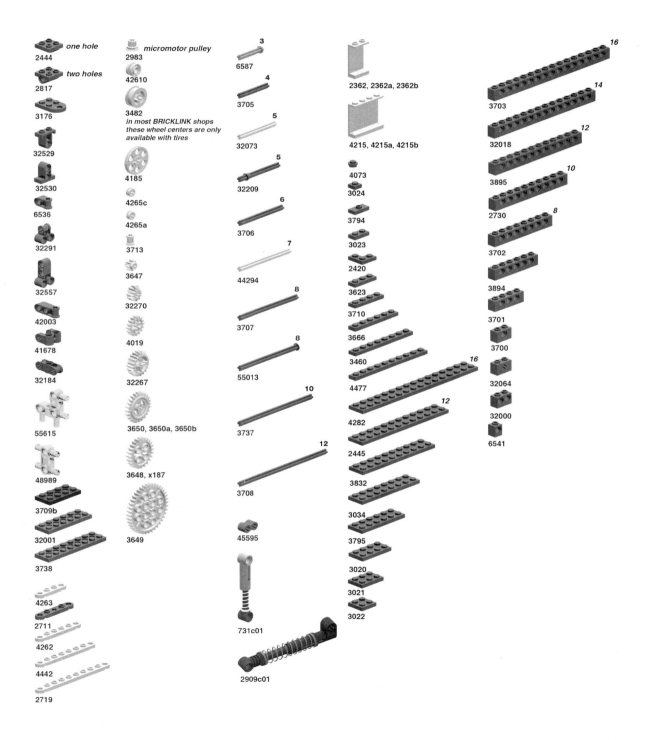

one hole
2444

two holes
2817

3176

32529

32530

6536

32291

32557

42003

41678

32184

55615

48989

3709b

32001

3738

4263

2711

4262

4442

2719

micromotor pulley
2983

42610

3482
in most BRICKLINK shops these wheel centers are only available with tires

4185

4265c

4265a

3713

3647

32270

4019

32267

3650, 3650a, 3650b

3648, x187

3649

3
6587

4
3705

5
32073

5
32209

6
3706

7
44294

8
3707

8
55013

10
3737

12
3708

45595

731c01

2909c01

2362, 2362a, 2362b

4215, 4215a, 4215b

4073

3024

3794

3023

2420

3623

3710

3666

3460

4477

16
4282

12
2445

3832

3034

3795

3020

3021

3022

16
3703

14
32018

12
3895

10
2730

8
3702

3894

3701

3700

32064

32000

6541

LDRAW: BUILD LEGO MODELS ON THE COMPUTER

LDraw (*http://www.ldraw.org/*) is an open standard with a digitized LEGO parts library, supplemented and updated by the community. LDraw is used by many programs to design LEGO models and generate building instructions.

However, there is no program for LDraw that verifies the mechanical functionality of a model. Hence, it is not possible to design functioning Technic models, such as my LEGO guns, solely on the computer.

To design the optimum color scheme for your models, create an LDraw file of your model. Because you can modify colors on the computer without having to change out actual pieces, color scheming is made much easier.

Here are some of the most important LDraw programs:

Mike's LEGO Cad (MLCad)

Using MLCad, you can build LEGO models with the computer. When you want to generate a construction manual for your model, you can use MLCad to separate the steps, from which the program LPub can generate building instructions.

My tips:

Parts can be pulled from the folder, on the left side of the screen, onto the working area by dragging and dropping. Because the MLCad-supplied folder structure is somewhat confusing, I suggest you fill your Favorites folder with the parts you use most often. To do that, you must manually list the part numbers in the *MLC_Favorites.txt* file.

The preselected icon color in the left parts list is black. I suggest changing it to light gray. The contours of the parts are then highlighted and are easier to identify.

When your model has more than 100 parts, you should split it into submodels, especially when you plan to write building instructions. LDraw files with submodels are called multipart documents and end in *.mpd*. My *Warbeast.mpd* file, for example, contains 30 submodels! All submodels can be dragged from the document folder.

MLCad has two modes: F2 activates view mode, and F3 activates edit mode. In view mode, you can cycle through the construction steps to preview the building instructions that LPub can generate.

MLCad's 3D view is only passable. You'll get a much better view running LDView alongside MLCad instead.

Keep your LDraw directory up-to-date. Install the unofficial parts library (*ldrawunf.zip*).

Newly installed parts must be added to the *parts.lst* file so that LDraw can find them. The program *mklist.exe* can do that automatically for you.

MLCad is the most used Lego design program, and it's easy and intuitive to use. Still, don't expect too much: There has not been a comprehensive update for a long time, and the interface seems out-of-date. Important standard functions such as undo (CTRL-Z) and the ability to work on multiple models at one time are absent.

LDView

LDView is a simple program that shows a rotatable 3D view of your LDraw model. It is an indispensable addition to MLCad for computer-aided LEGO designing.

Over the years, LDView has been expanded to include several extremely useful features:

Save Snapshot   Generates a screenshot of the current display. The resolution is freely selectable.

Export   Creates a *.pov* file. The freeware rendering program POV-Ray can open this file and generate a high-quality image of your model with hard shadows, reflections, and other effects.

Check for Library Updates . . .   Checks the online LDraw database, and when new LEGO parts are found, automatically installs them.

Polling   Causes LDView to automatically update its view as soon as the open LDraw file is changed. This function is vital when you use LView with MLCad.

Parts List . . .   Exports an HTML page with graphical renderings and part numbers of all parts used in the currently open model. This is convenient when publishing models on the Internet.

Preferences   Includes countless options for model depiction, such as shading, perspective, grids, anti-aliasing, and much more.

Press SHIFT to reposition the light source using the mouse.

LPub

Use LPub to produce sophisticated step-by-step construction documentation. LPub imports LDraw files and generates the necessary pictures with construction steps and part lists. Earlier LPub versions (1 through 3) required you to give specific instructions for where and how the pictures should be rendered to be written in the LDraw file (by using, for example, MLCad). Then

a layout program was needed to combine the pictures and construction steps.

In contrast, the version current as of this writing has a built-in editor with which you can lay out, and export, the construction manual.

I still prefer version 2.3.0.10 of LPub (used for this book), since it implements a software renderer. I like software renderers because they generate striking, almost photorealistic pictures with multiple light sources, lustrous surfaces, and hard shadows. The current version uses the hardware renderer of the video card (OpenGL), which is much faster but delivers artless pictures. The old LPub took six hours to render the Warbeast building instructions, while the new version probably would have taken only five minutes.

Regardless of the LPub version you use, each is relatively complicated and takes some time to learn.

BrickStore

The BrickStore program was written for BrickLink sellers to help update their LEGO part offers.

BrickStore is also a practical tool for LEGO hobbyists. It can import LDraw files and calculate the final weight of a model, number of parts needed, and roughly how expensive it would be to order the parts from BrickLink. Furthermore, it checks to see whether the model parts in the selected colors even exist. For example, dark bluish gray Technic bricks are available in all lengths except 8 and 10 simply because these bricks have never shown up in any LEGO set.

BrickStore can also export BrickLink wanted lists, which can be helpful when you want to buy the model parts.

PARABELLA

★ ★ ★ ★ ★

A SEMIAUTOMATIC RUBBER-BAND PISTOL

rubber band
catch

barrel

trigger
guard

AMMO: *15 rubber bands*
LENGTH: *26.42 cm*
WIDTH : *4.01 cm*
HEIGHT: *16.95 cm*
WEIGHT: *200 g*
PARTS: *193*
FIRE POWER: ★ ★ ★ ★ ★
LEVEL: *Beginner*

Ammunition for the Parabella:
a 55 mm diameter rubber band

The Parabella, fully loaded and ready to fire!

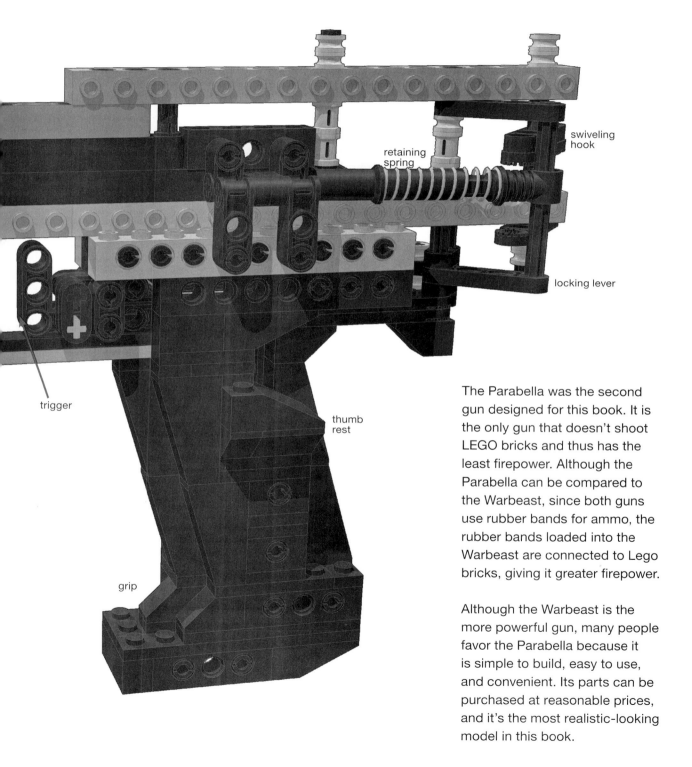

retaining
spring

swiveling
hook

locking lever

trigger

thumb
rest

grip

The Parabella was the second gun designed for this book. It is the only gun that doesn't shoot LEGO bricks and thus has the least firepower. Although the Parabella can be compared to the Warbeast, since both guns use rubber bands for ammo, the rubber bands loaded into the Warbeast are connected to Lego bricks, giving it greater firepower.

Although the Warbeast is the more powerful gun, many people favor the Parabella because it is simple to build, easy to use, and convenient. Its parts can be purchased at reasonable prices, and it's the most realistic-looking model in this book.

HOW IT WORKS

1

loaded
rubber bands

trigger

retaining
spring

locking
lever

swiveling
hook

2

3

In the Parabella's initial position, two rubber bands are pulled taut over the swiveling hook, creating strong, counterclockwise torque on the hook. The hook cannot move because it is held in position by the locking lever (marked above by the red circle).

The Parabella is ready to be fired.

When the trigger is pulled, the locking lever flips to the side, thereby freeing the swiveling hook and allowing it to turn. One of the loaded rubber bands is released, catapulting forward as a result of its own tension. After a quarter turn (90°), the swiveling hook is stopped once more by the locking lever.

With each pull of the trigger, one rubber band is released in this manner.

When the trigger is released, the retaining spring returns the trigger and locking lever to their initial positions, allowing the swiveling hook to make the final quarter turn back to its original position. The weapon is ready to be fired again.

Depending on the thickness of your rubber bands, the Parabella can be loaded with up to 15 rounds.

BILL OF MATERIALS

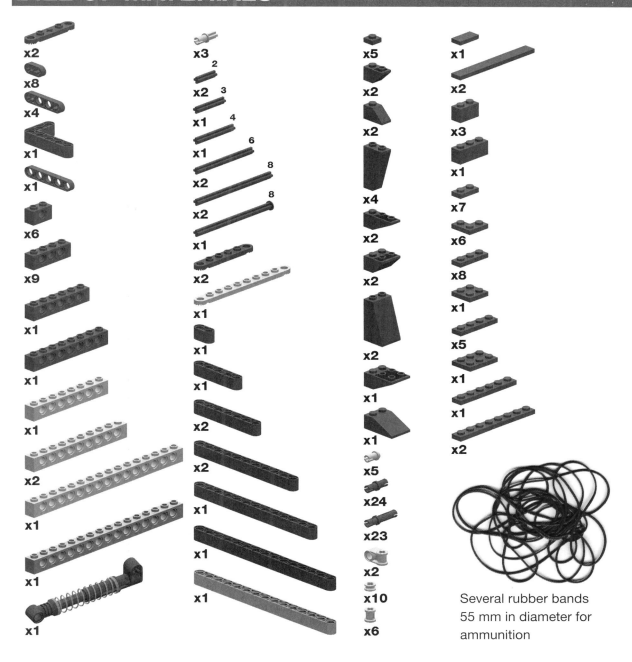

x2

x8

x4

x1

x1

x6

x9

x1

x1

x1

x2

x1

x1

x1

x3

x2 **2**

x2 **3**

x1 **4**

x1 **6**

x2 **8**

x2 **8**

x1

x2

x1

x1

x1

x2

x2

x1

x1

x1

x5

x2

x2

x4

x2

x2

x1

x2

x1

x1

x5

x24

x23

x2

x10

x6

x1

x2

x3

x1

x7

x6

x8

x1

x5

x1

x1

x2

Several rubber bands
55 mm in diameter for
ammunition

GRIP

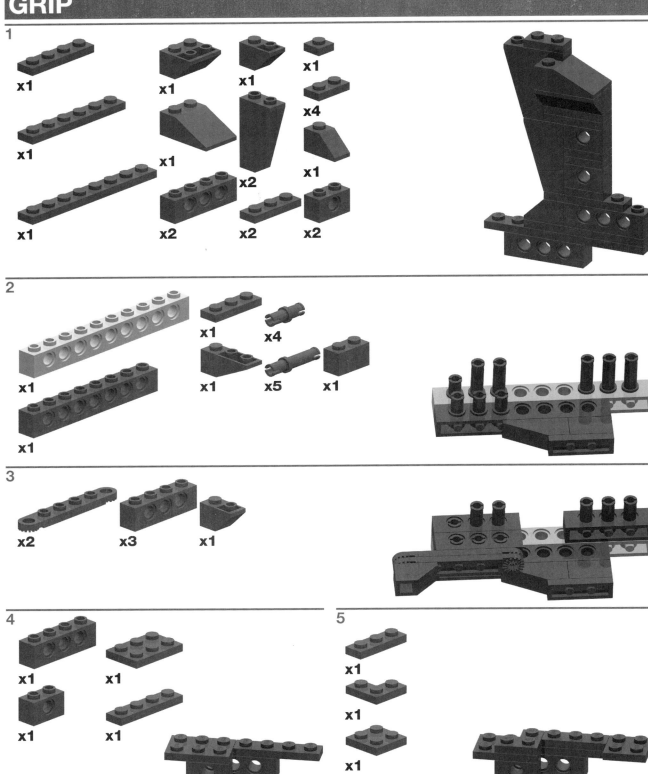

1

x1 x1 x1 x1

x1 x1 x4

x1 x1 x1

x1 x2 x2 x2

2

x1 x1 x4

x1 x5 x1

x1

3

x2 x3 x1

4

x1 x1

x1 x1

5

x1

x1

x1

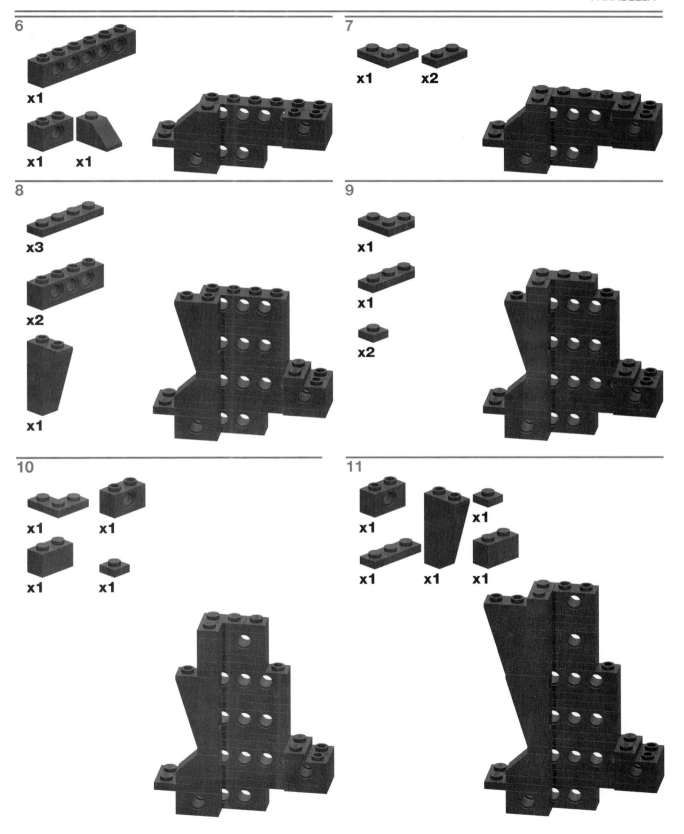

6

x1

x1 x1

7

x1 x2

8

x3

x2

x1

9

x1

x1

x2

10

x1 x1

x1 x1

11

x1 x1

x1 x1 x1

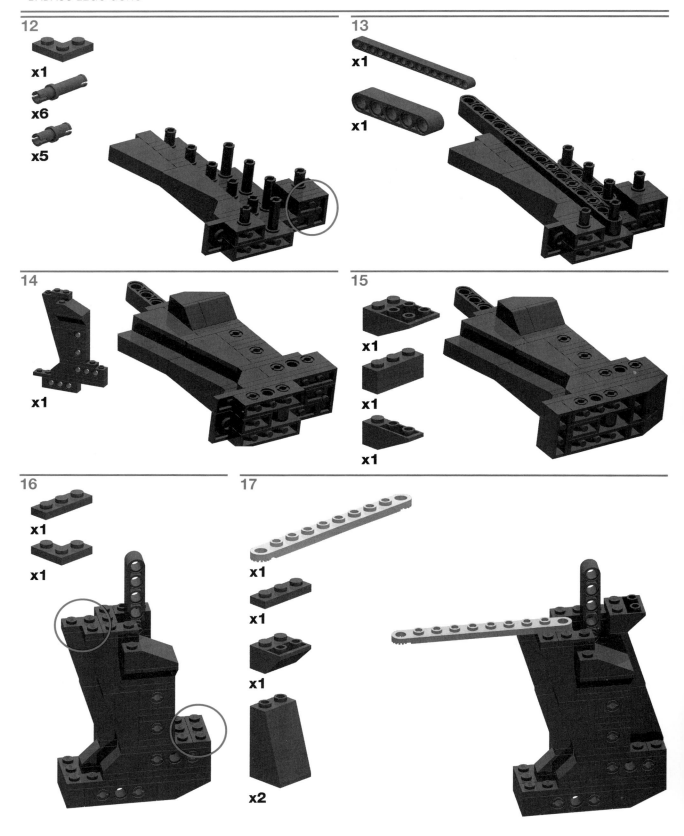

12

x1

x6

x5

13

x1

x1

14

x1

15

x1

x1

x1

16

x1

x1

17

x1

x1

x1

x2

18

x1

x1

x1

19

x1

x1

20

x1

21

x1

22

x1

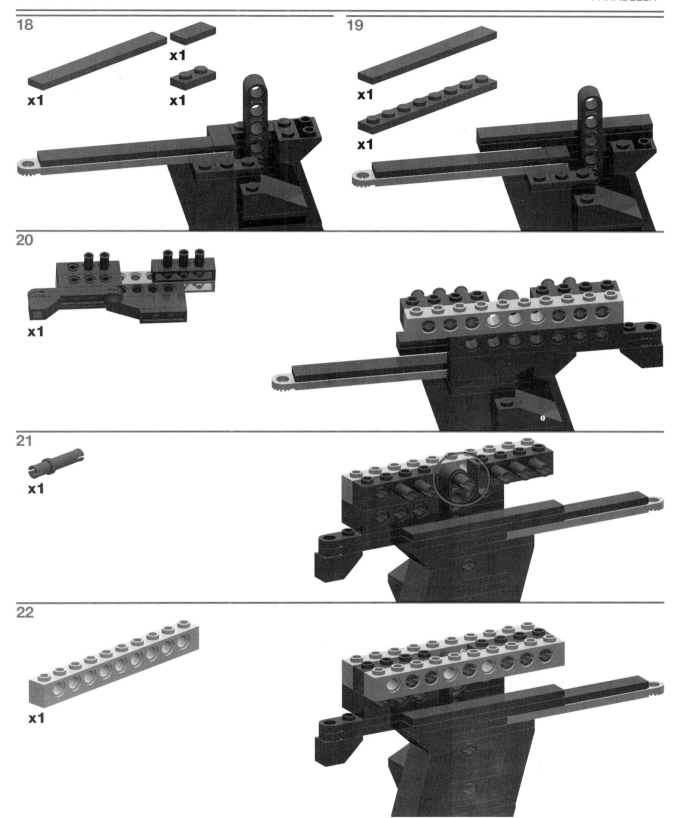

BARREL

1

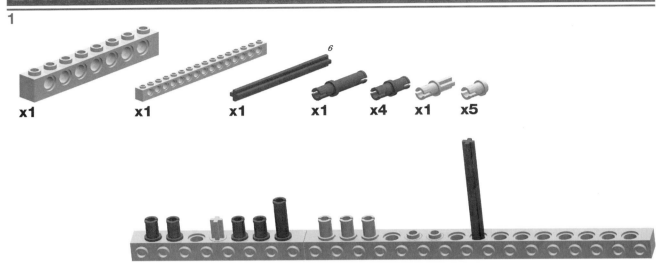

x1   x1   x1   x1   x4   x1   x5

2

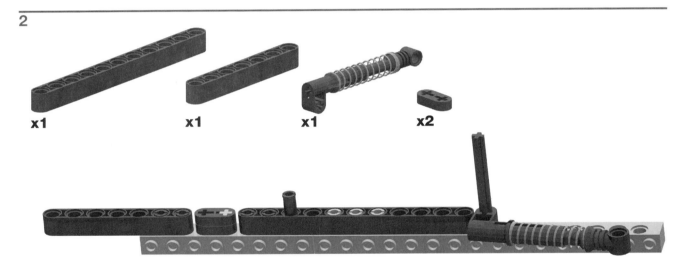

x1   x1   x1   x2

3

x5   x1   x2

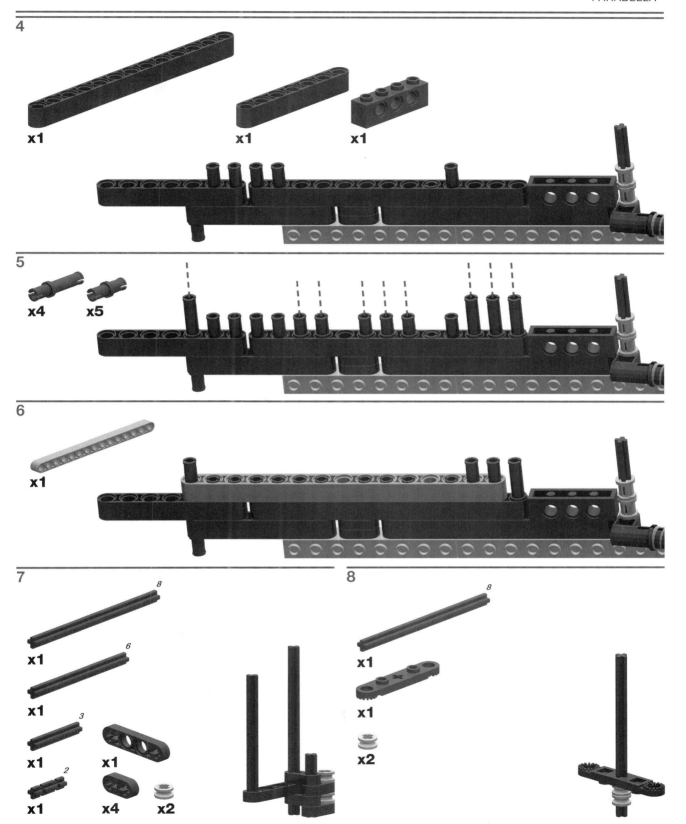

4

x1   x1   x1

5

x4   x5

6

x1

7

8

x1

6

x1

3

x1   x1

2

x1   x4   x2

8

8

x1

x1

x2

9

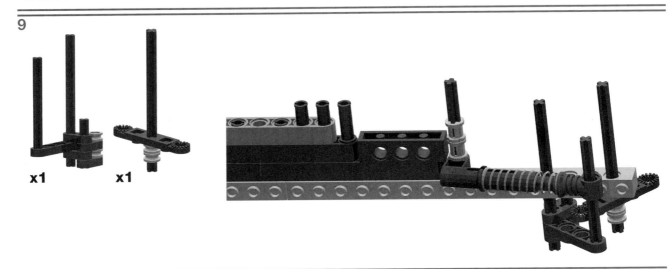

x1 x1

10

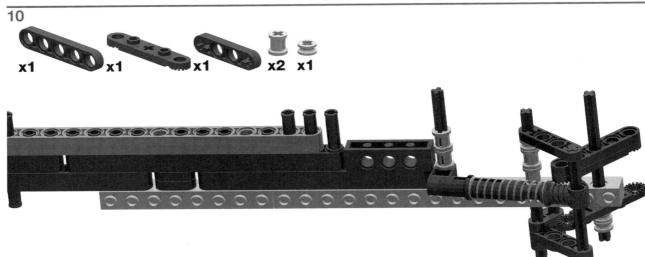

x1 x1 x1 x2 x1

11

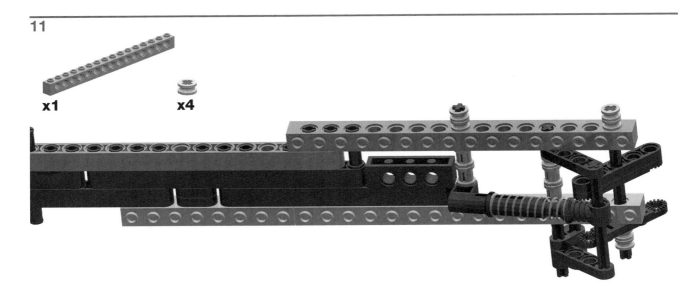

x1 x4

FINAL ASSEMBLY

1

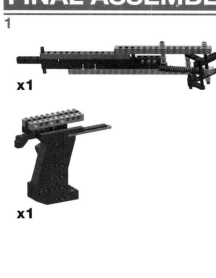

x1

x1

2

x1　x1　x1

x1　x1　x2

3

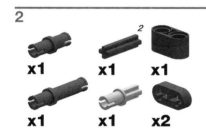

x1　x2

x1

x1

4

x1

x1

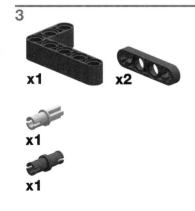

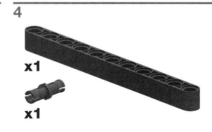

5

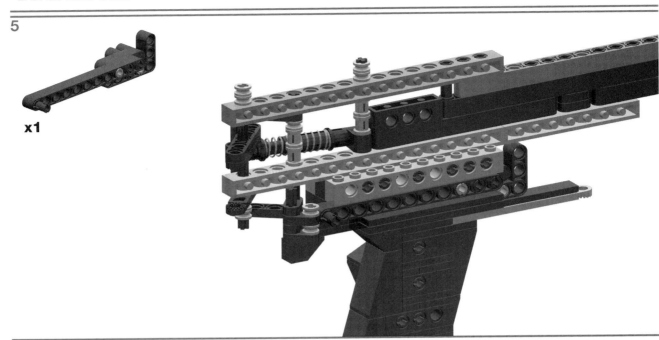

x1

6

4

x1 **x2** **x2** **x2**

7

8

x1 **x1**

x1

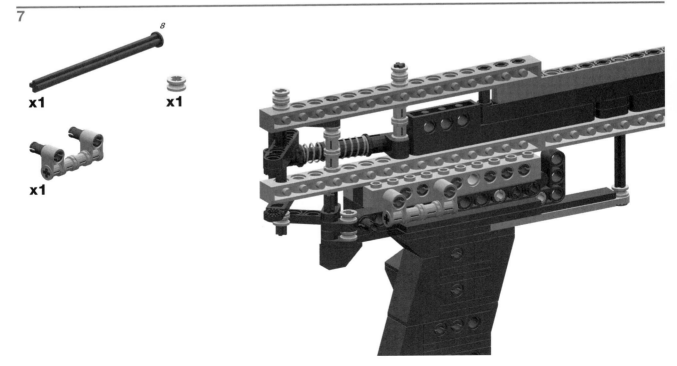

8

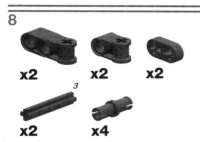

x2   x2   x2

x2   x4

9

x1

LOADING THE GUN

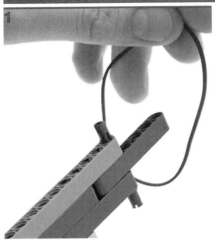

1

Loop the rubber band over the catch at the front of the gun.

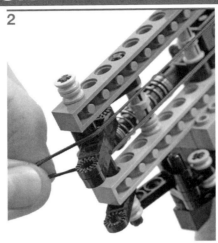

2

Pull the rubber band, and hook it around the swiveling hook.

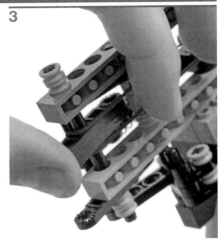

3

Turn the swiveling hook clockwise.

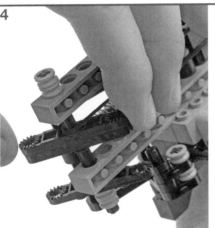

4

5

Continue to turn the swiveling hook until it completes a 180° rotation.

Repeat steps 1–4 to load more rubber bands.

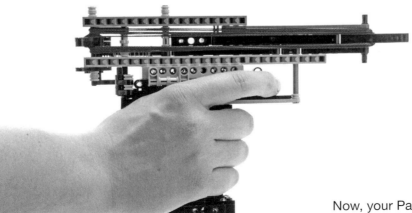

Now, your Parabella is fully loaded!

LILIPUTT

★ ★ ★ ★ ★

A SEMIAUTOMATIC, DOUBLE-ACTION PISTOL

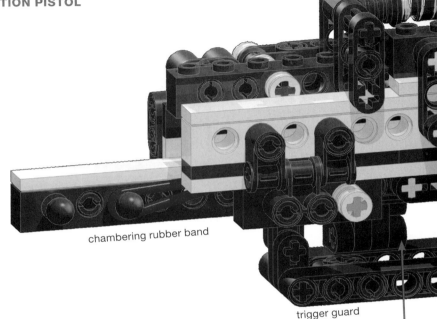

spring

barrel

chambering rubber band

trigger guard

trigger

AMMO: *nine 2×1 bricks*
LENGTH: *25.04 cm*
WIDTH: *6.65 cm*
HEIGHT: *15.88 cm*
WEIGHT: *250 g*
PARTS: *321*
FIRE POWER: ★ ★ ★ ★ ★
LEVEL: *Intermediate*

The Liliputt projectile

The Liliputt with lid opened for reloading

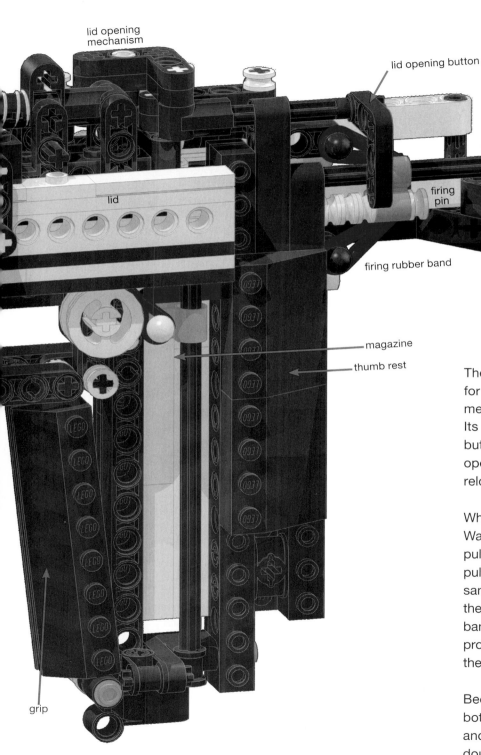

lid opening
mechanism

lid opening button

lid

firing
pin

firing rubber band

magazine

thumb rest

grip

The first model designed for this book, the Liliputt is mechanically quite elaborate. Its special feature is a push-button release lid that, when open, frees the magazine for reloading.

Whereas the rubber bands in the Warbeast and the Parabella are pulled taut when loaded and the pullback of the slide does the same for the Mini-Thriller and the Thriller Advanced, the rubber band that fires the Liliputt's projectiles remains slack until the trigger is pulled.

Because pulling the trigger both stretches the rubber band and fires the weapon, this is a double-action pistol.

LILIPUTT DESIGN HISTORY

The goal of the Liliputt was to design a semiautomatic pistol that fires 2×1 LEGO bricks.

1

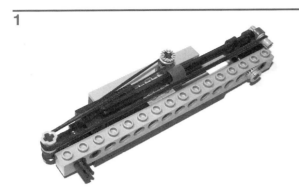

First I developed a magazine for the 2×1 bricks.

2

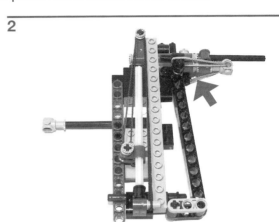

Then I designed the shooting mechanism.

3

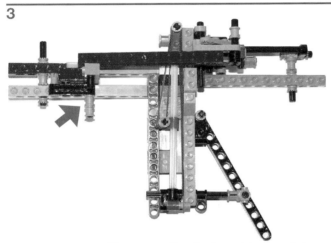

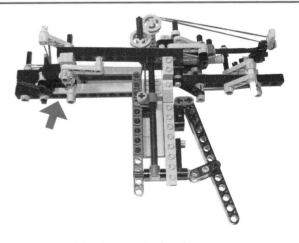

These are the first working prototypes, complete with trigger mechanisms.

4

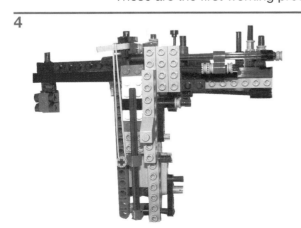

This is the first perfectly functioning prototype.

5

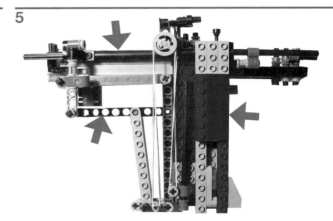

Here, I've added the barrel, handle, and trigger guard.

6

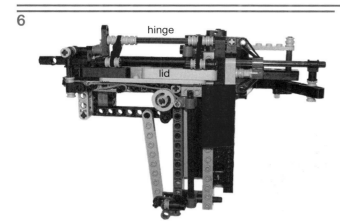

To open the magazine for reloading, I laid out the left side of the barrel as a kind of lid. My initial idea was to put its hinge along the top, parallel to the barrel, with the lid opening sideways and up.

7

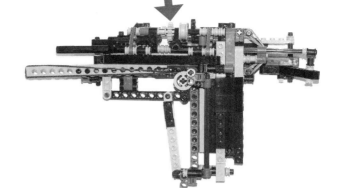

The mechanism shown here was to be a catch for the lid.

8

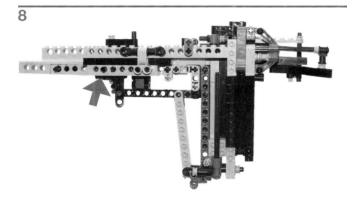
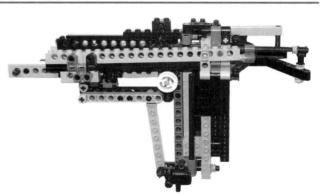

The side-opening lid didn't work particularly well, so I switched to a front-hinged, top-opening one.

9

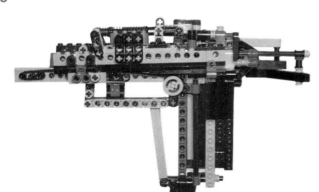

The lid was to be opened with the right thumb. I designed a complicated mechanism, consisting of a spring and several levers, to accomplish this.

10 FINAL VERSION

Only once all mechanical problems were solved did I decide on the final color scheme, using LDraw.

HOW IT WORKS: THE LID MECHANISM

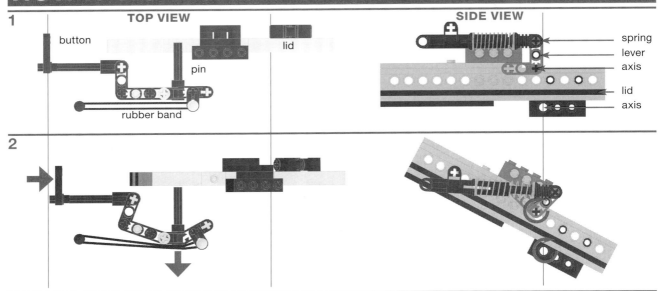

TOP VIEW

button

pin

lid

rubber band

SIDE VIEW

spring

lever

axis

lid

axis

HOW IT WORKS: THE FIRING MECHANISM

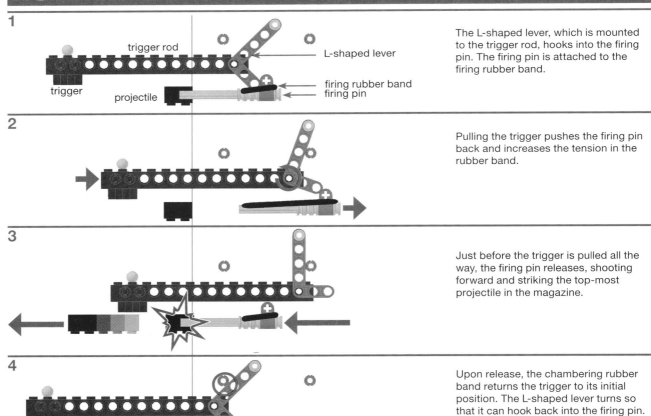

1

trigger rod

L-shaped lever

firing rubber band

firing pin

trigger

projectile

The L-shaped lever, which is mounted to the trigger rod, hooks into the firing pin. The firing pin is attached to the firing rubber band.

2

Pulling the trigger pushes the firing pin back and increases the tension in the rubber band.

3

Just before the trigger is pulled all the way, the firing pin releases, shooting forward and striking the top-most projectile in the magazine.

4

Upon release, the chambering rubber band returns the trigger to its initial position. The L-shaped lever turns so that it can hook back into the firing pin.

BILL OF MATERIALS

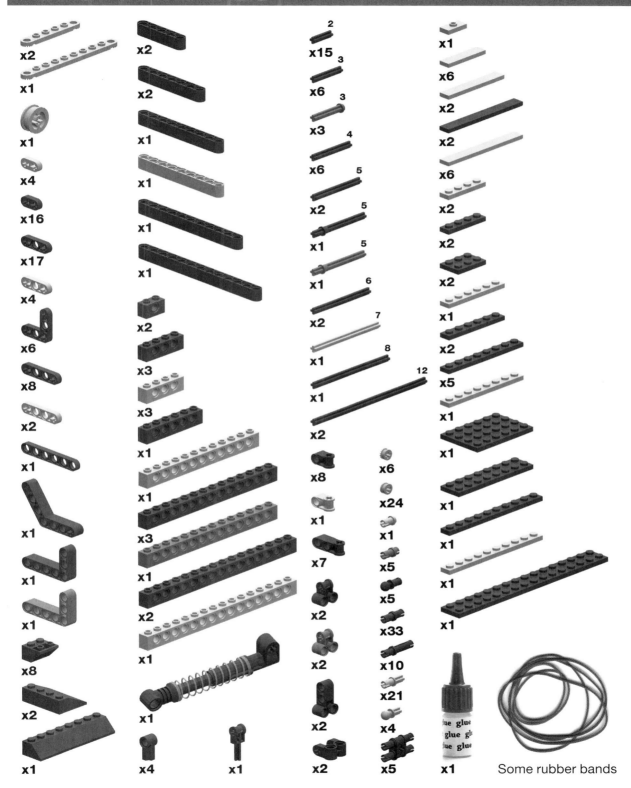

x2

x1

x1

x4

x16

x17

x4

x6

x8

x2

x1

x1

x1

x1

x8

x2

x1

x2

x2

x1

x1

x2

x3

x3

x1

x1

x3

x1

x2

x1

x1

x4

x1

2
x15

3
x6

3
x3

4
x6

5
x2

5
x1

5
x1

6
x2

7
x1

8
x1

12
x2

x8

x1

x7

x2

x2

x2

x2

x6

x24

x1

x5

x5

x33

x10

x21

x4

x5

x1

x6

x2

x2

x6

x2

x2

x2

x1

x2

x5

x1

x1

x1

x1

x1

x1

x1

x1

x1

Some rubber bands

35

BARREL BOTTOM

1

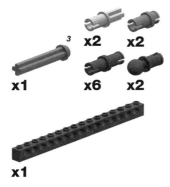

3
x1 **x2** **x2**
x6 **x2**

x1

2

x1

x1 **x1**

x1

3

x2

x1

4

x3

x1

5

x1

x2

x1

6

x1

x3

x2

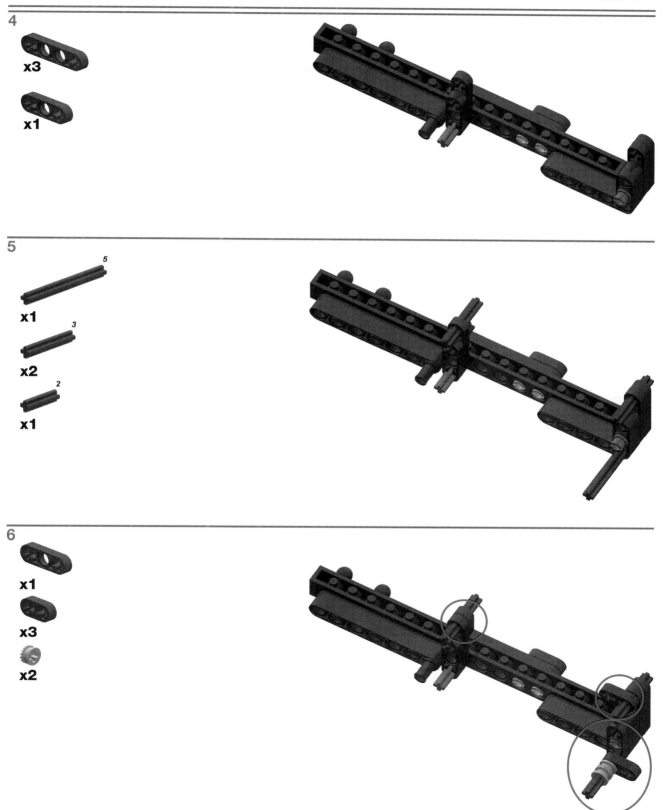

7

x1
5

x1
3

8

x2

x1

x1

9

x1

x1

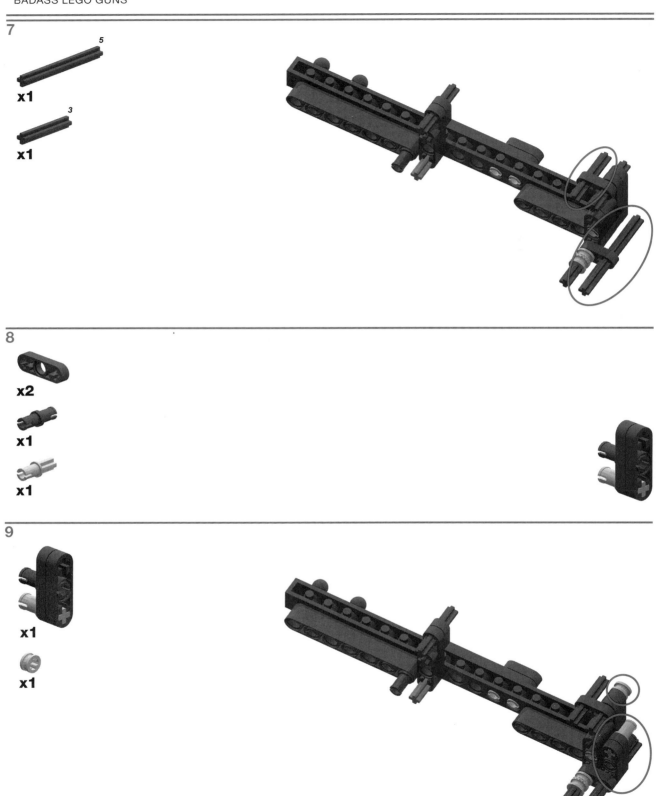

10

x1

11

x1

x2

12

x1

x1

x1

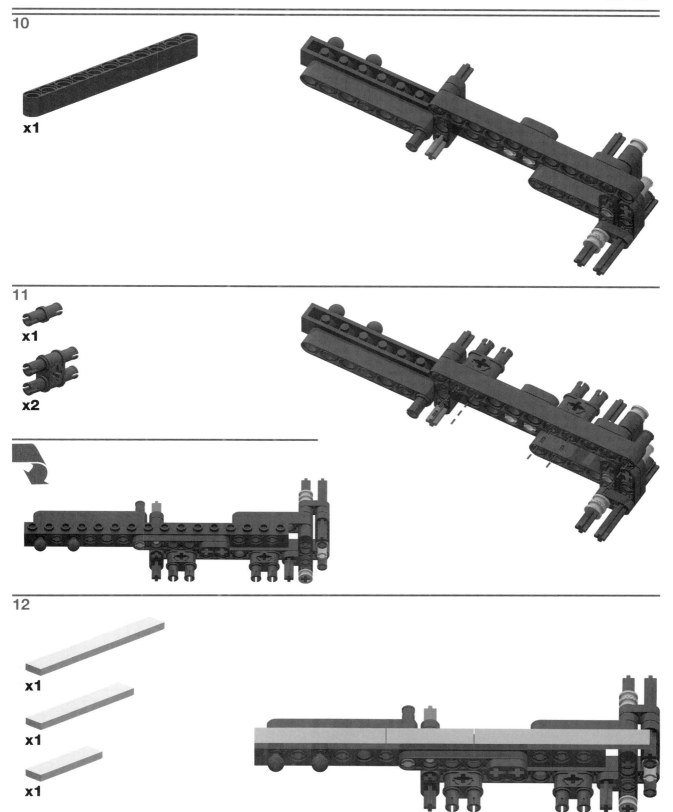

RIGHT SECTION

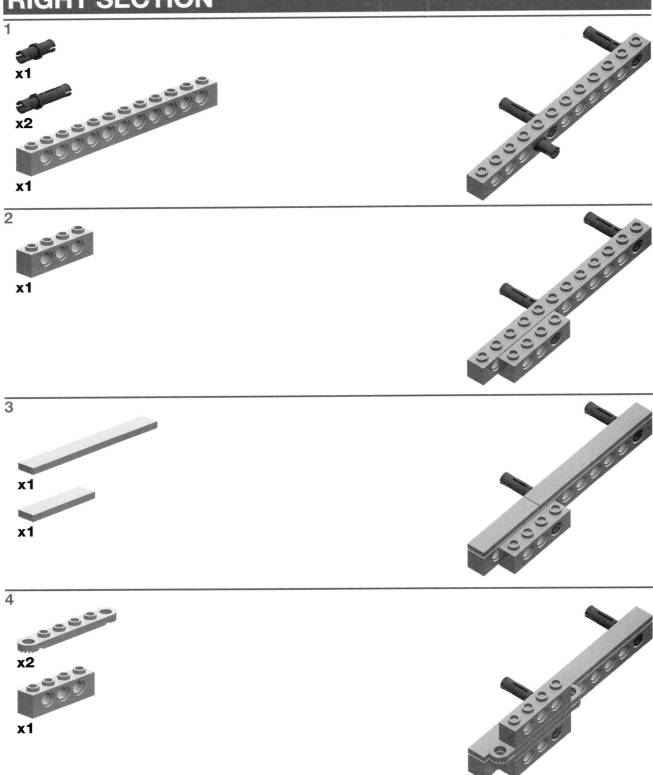

1

x1

x2

x1

2

x1

3

x1

x1

4

x2

x1

5

x2

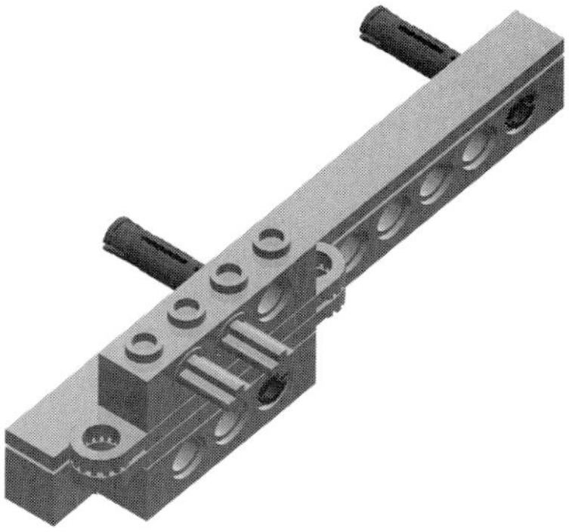

6

x2

x1

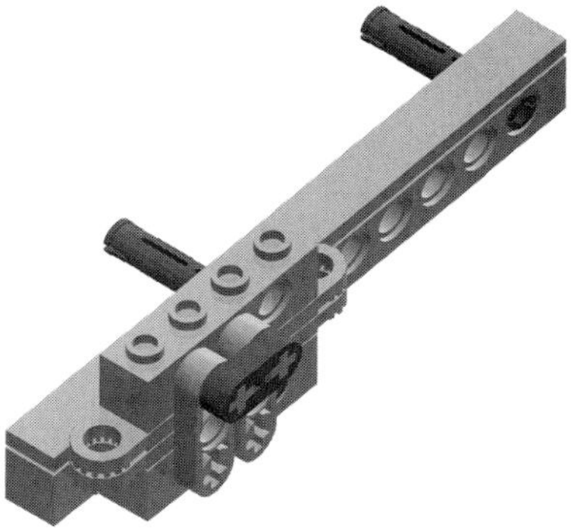

7

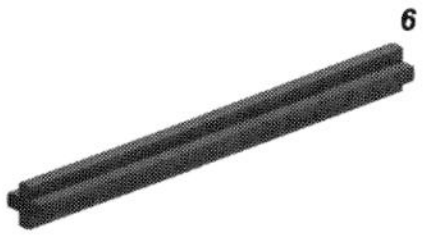

6

x2

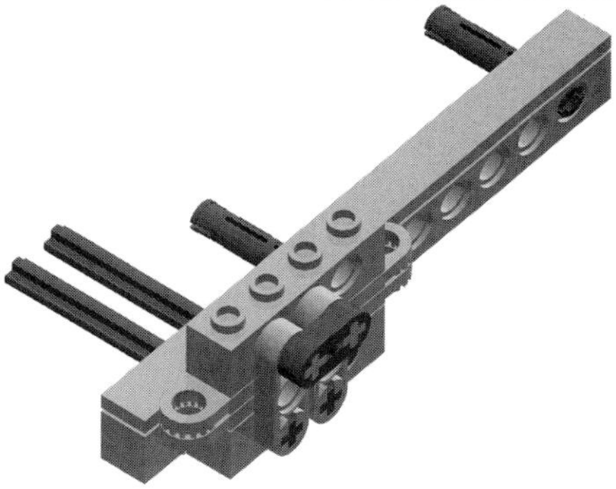

8

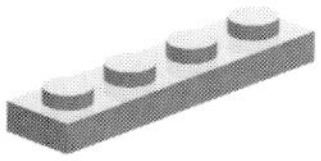

x2

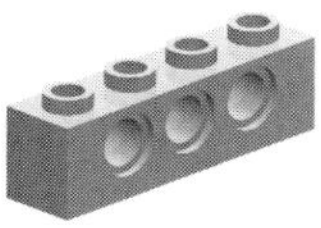

x1

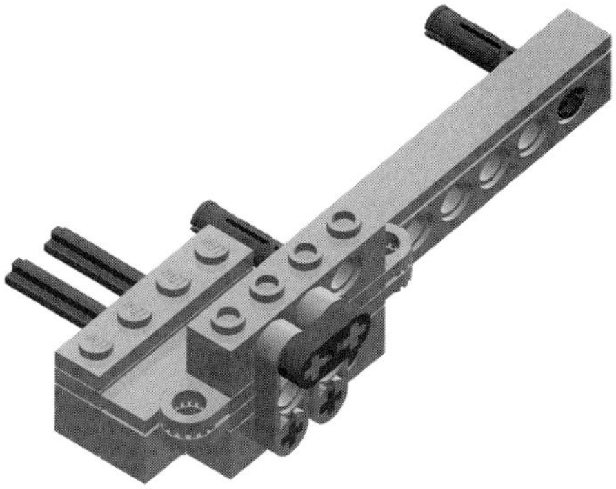

BARREL: RIGHT SECTION

1

x2

x5

x2

2

x2

x1

3

4

x3

4

x1

x4

x3

5

x1

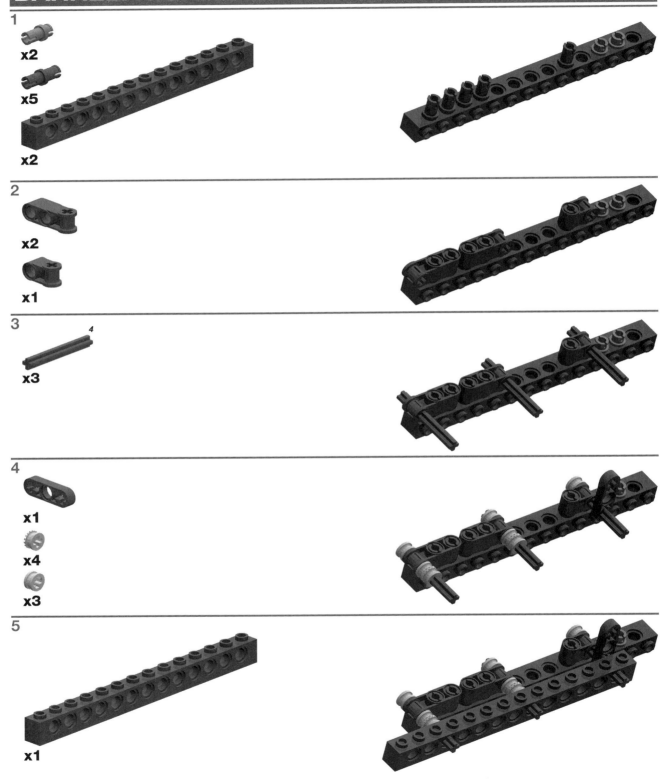

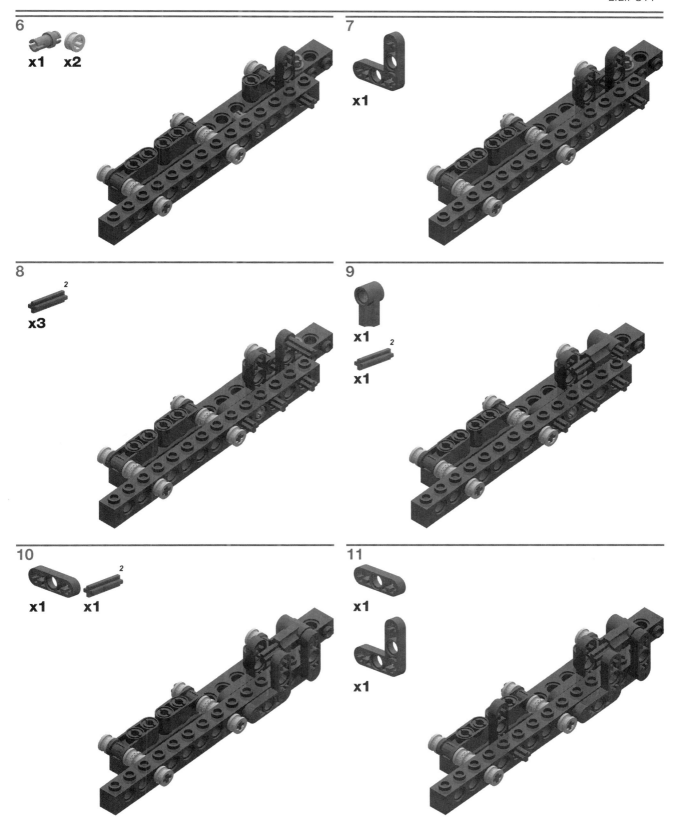

6 x1 x2

7 x1

8 x3

9 x1 x1

10 x1 x1

11 x1 x1

12

x1

x1

13

x1

14

x1

x2

x1

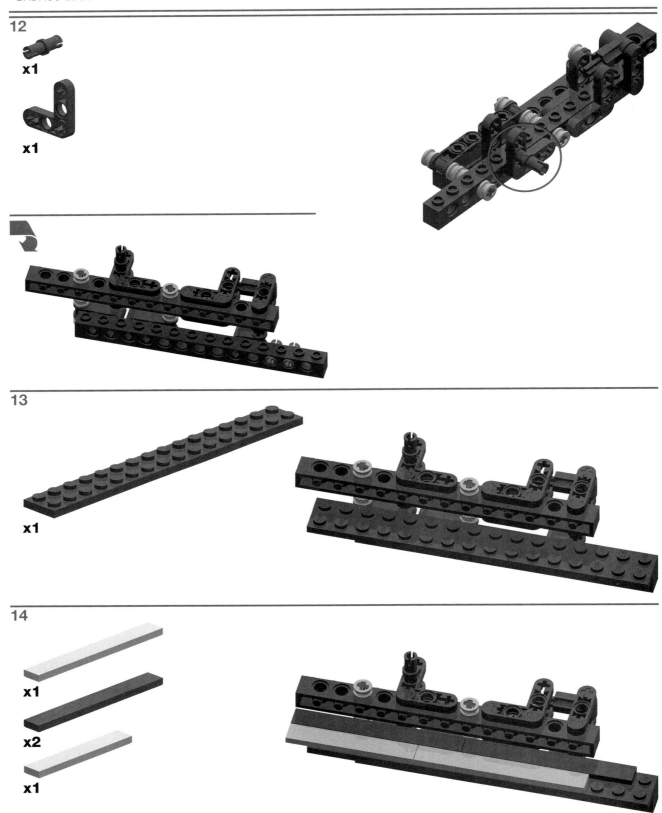

LID

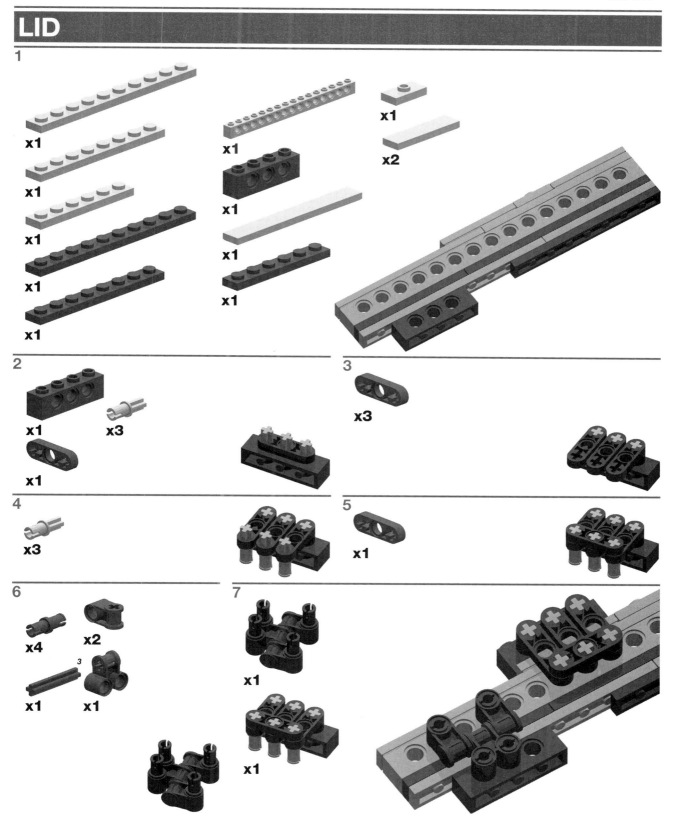

1

x1

x1

x1

x1

x1

x1

x1

x1

x2

x1

x1

2

x1   x3

x1

3

x3

4

x3

5

x1

6

x4   x2

3

x1   x1

7

x1

x1

FINAL ASSEMBLY

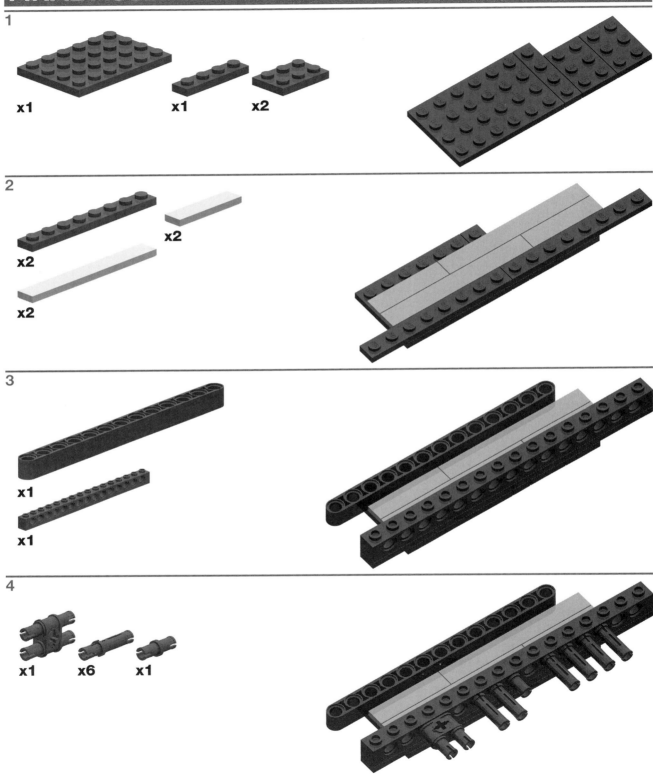

1

x1   x1   x2

2

x2
x2
x2

3

x1
x1

4

x1   x6   x1

5

x1 x1 x1

6

x1

x1

7

x2

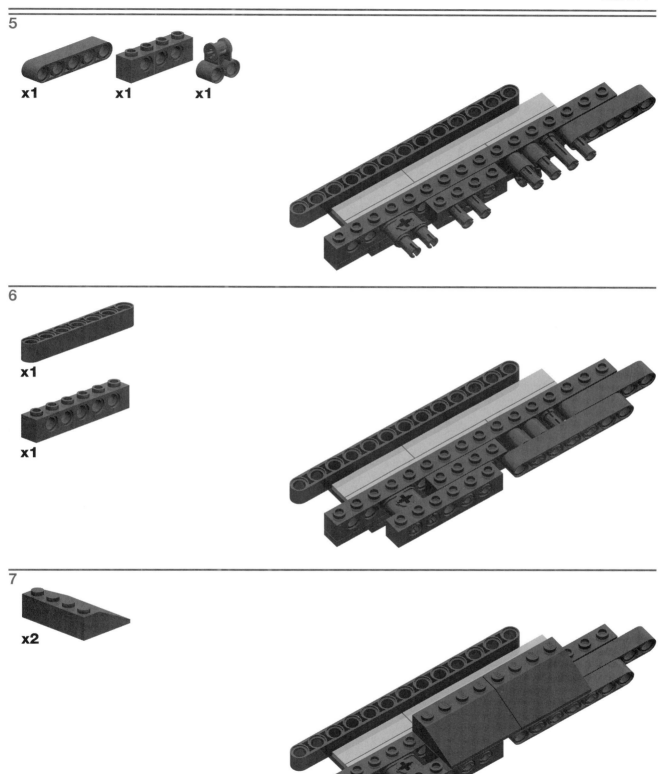

8

x1

9

x1

x1

x1

10

x4

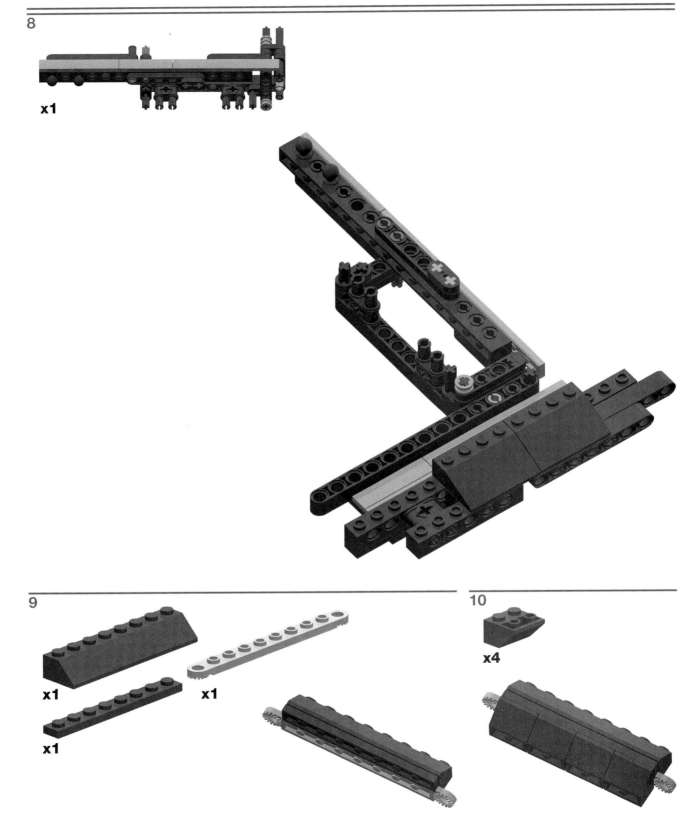

11

x1

12

x1

x1

13

x1

x1

x1

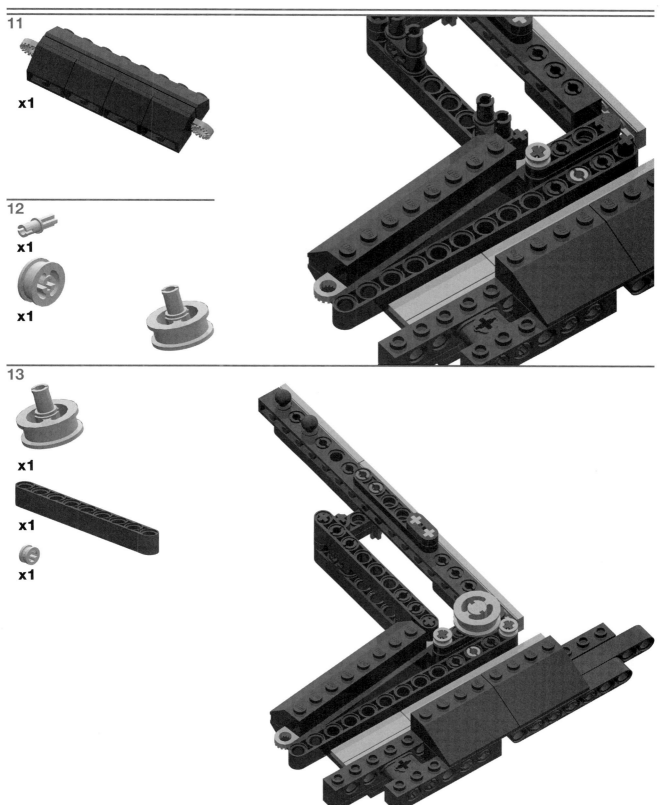

14

x1

x1

x3

15

glue glue
glue gl
lue glue

x1

16

x1

17

x1

18

x1

x1 x1 x2

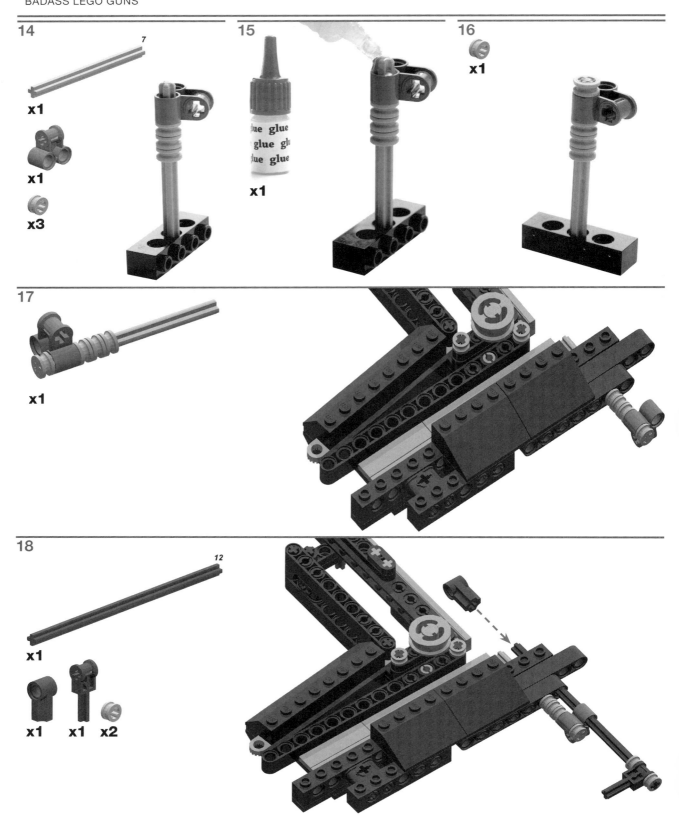

19

x1

3

x1 x1

x1 x1 *2*

20

x1

2

x1

21

x1

2

x1

22

x1

23

x1

24

12

x1

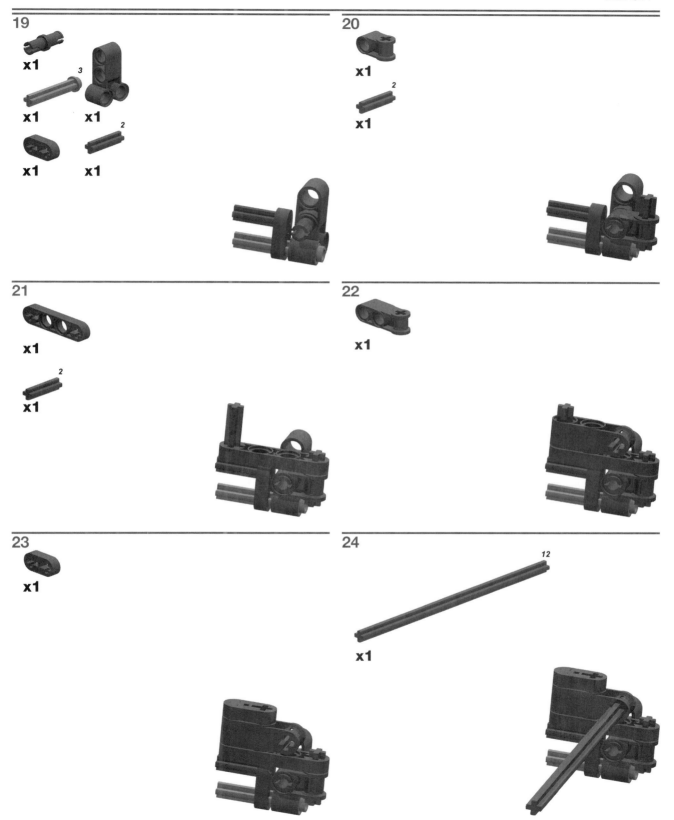

25

26

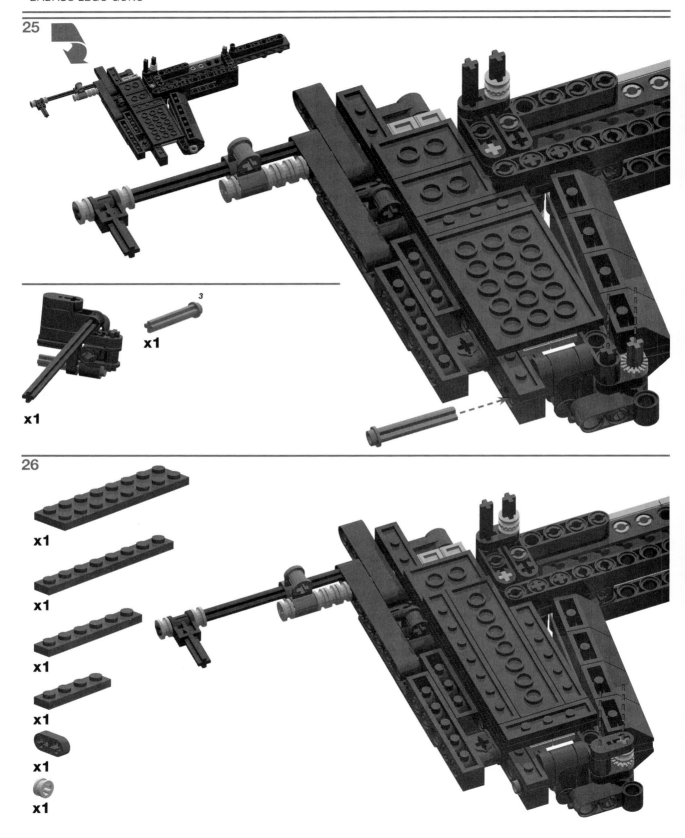

27

x1

x4

28

x1 x2 x1 x1 x1

29

x1

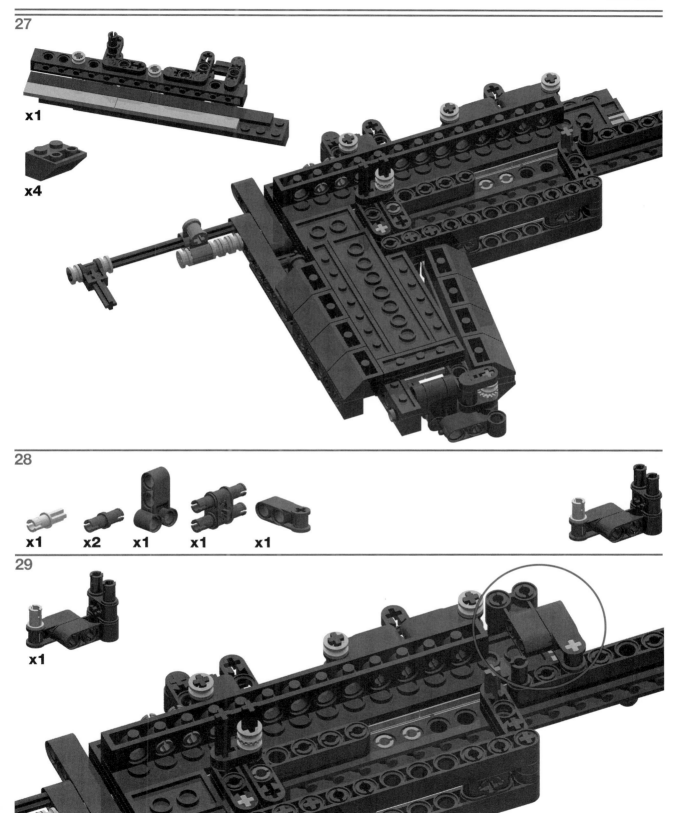

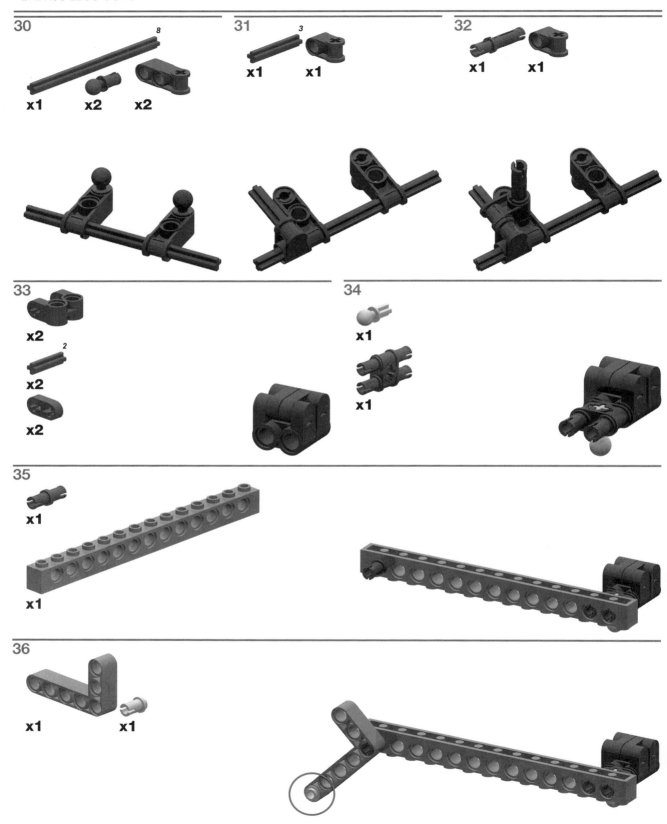

30
x1 x2 x2

31
x1 x1

32
x1 x1

33
x2
x2
x2

34
x1
x1

35
x1
x1

36
x1 x1

37

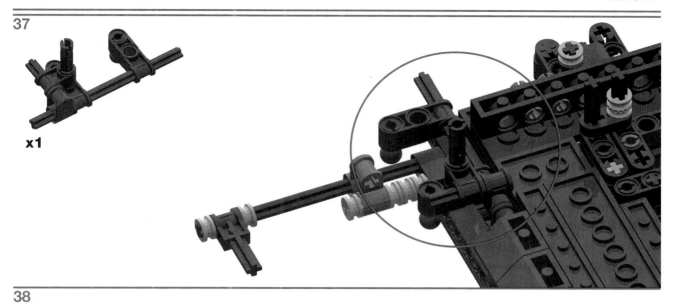

x1

38

x1

39

x1

40

x1 x2 x4

41

x1

42

x1

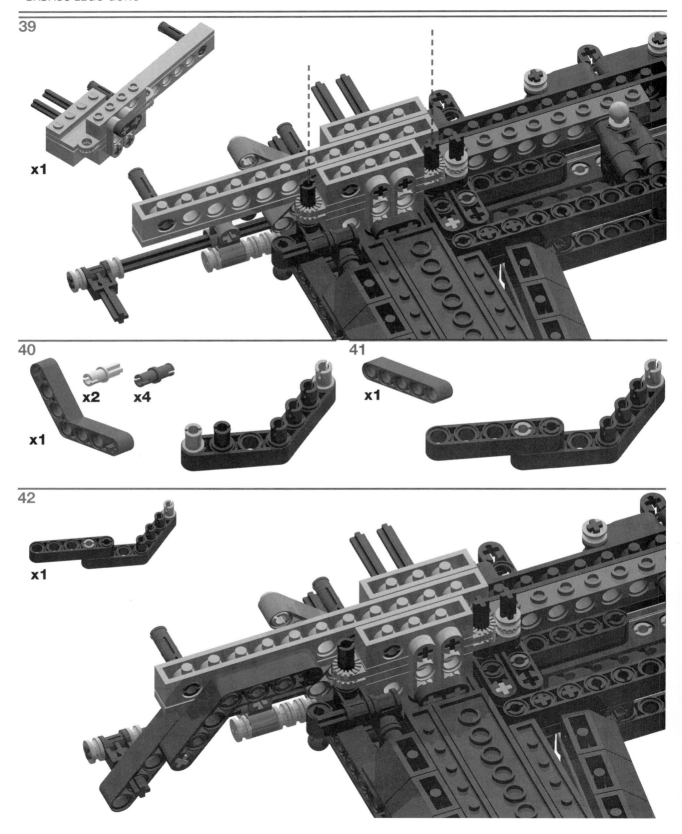

43

x1

x1 ₂

44

x2

x2

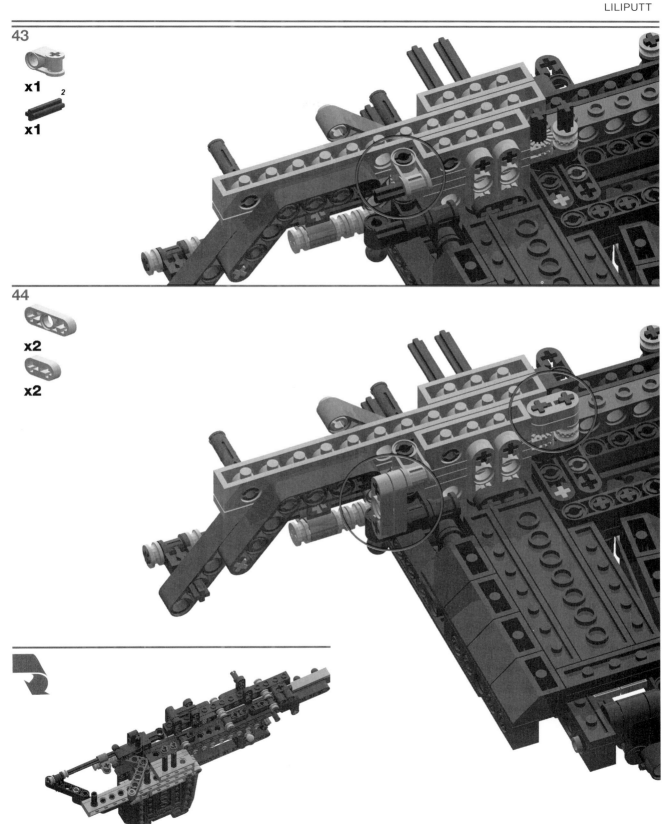

45

x1

x1

46

x2

x2

47

x2   **x1**   **x1**   **x1** `4`   **x1** `3`

x1   **x3**   **x1**

48

x1

x1

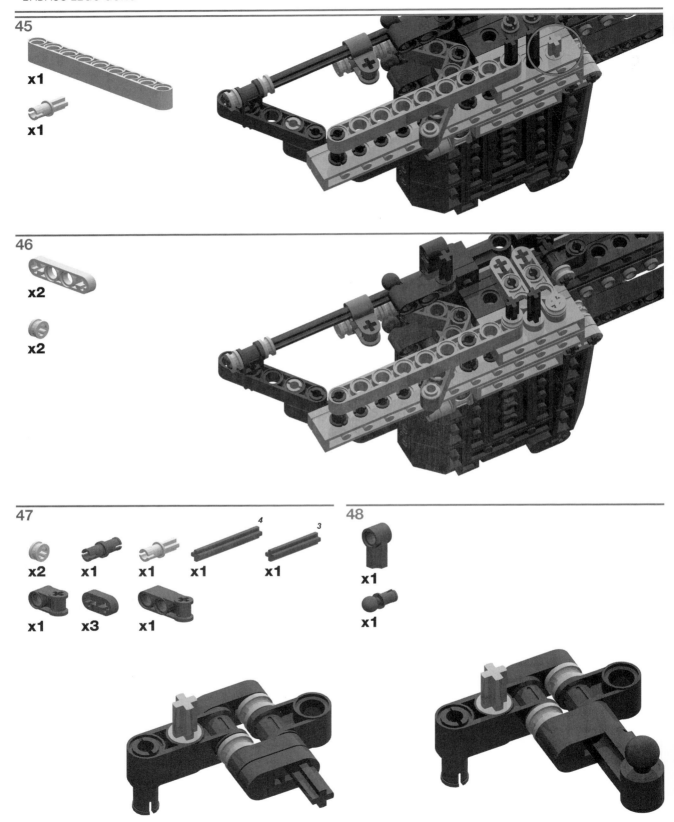

49

x1

50

x2

51

x2

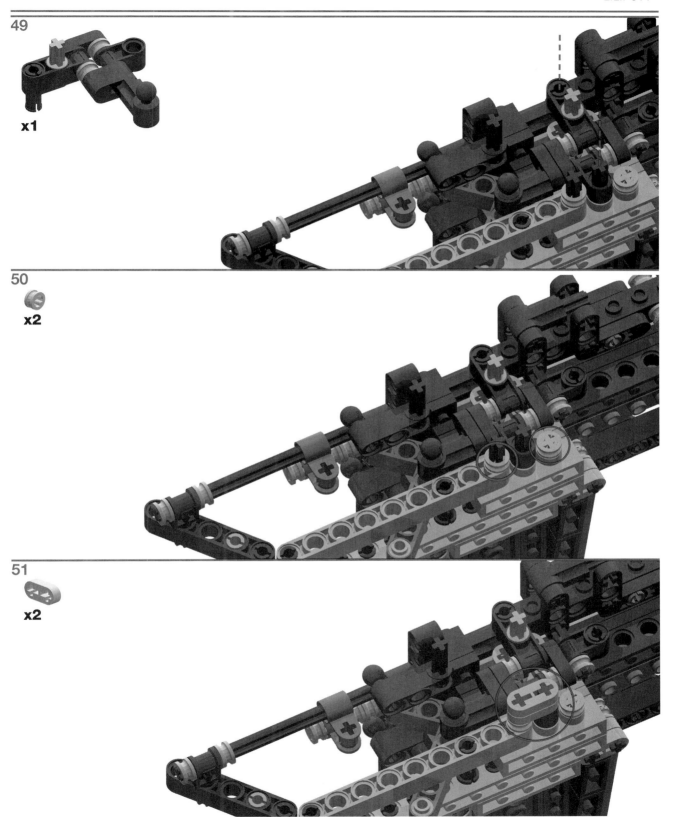

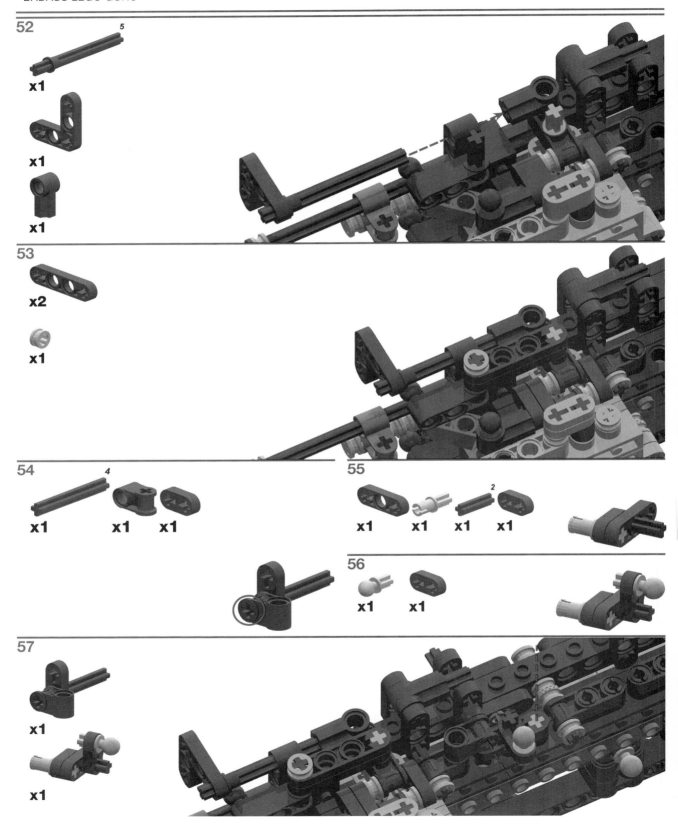

52

x1

x1

x1

53

x2

x1

54

x1   x1   x1

55

x1   x1   x1   x1

56

x1   x1

57

x1

x1

58

x1

Be sure this rubber band
is not too taut. Figure out
what works best.

59

x2

60

2

x2

x2

61

x2

62

x1   x1

63

x1

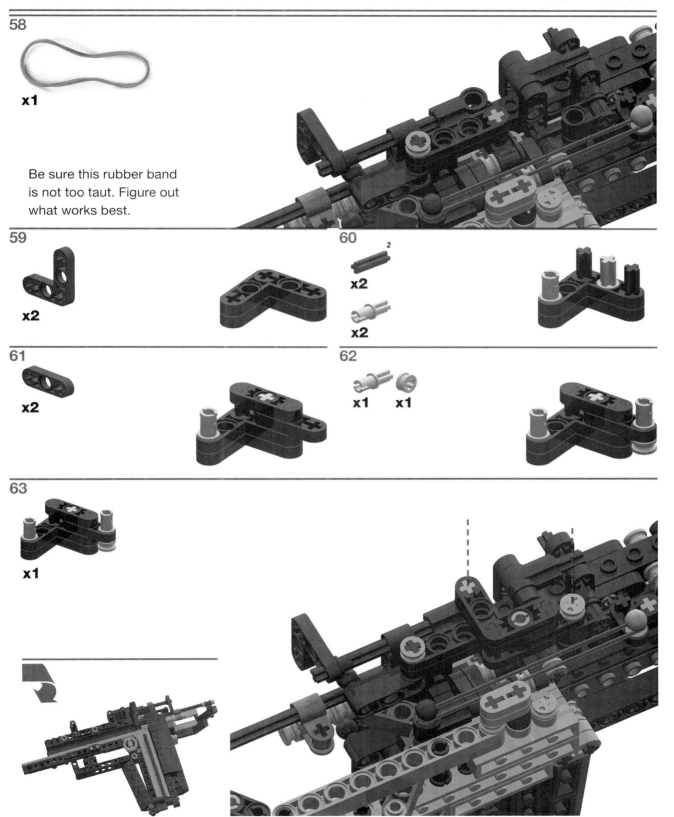

64

x2

x1

65

x1

x1

66

x1

x1

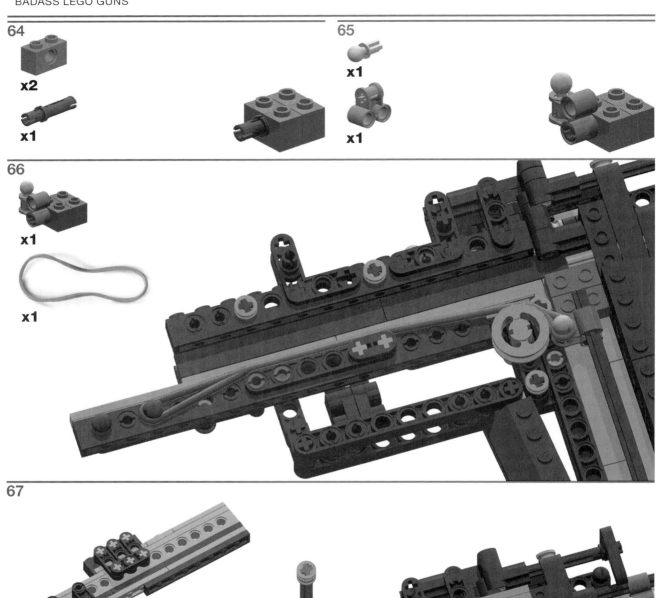

67

x1

5

x1

x2

68

x2

69

x2

70

x1

71

x1

72

x1

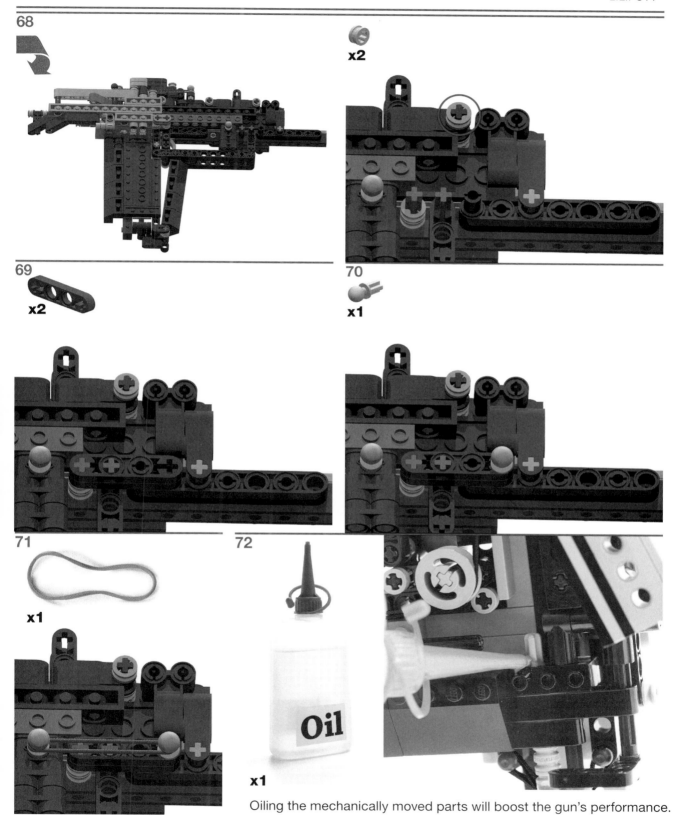

Oiling the mechanically moved parts will boost the gun's performance.

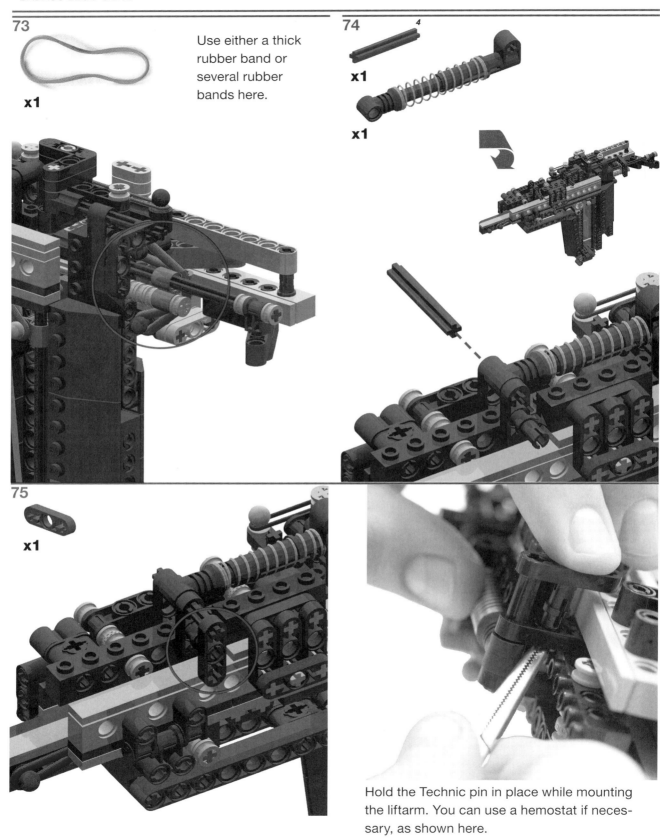

73

x1

Use either a thick
rubber band or
several rubber
bands here.

74

4

x1

x1

75

x1

Hold the Technic pin in place while mounting
the liftarm. You can use a hemostat if necessary, as shown here.

LOADING THE GUN

1

Press the open button with your thumb.

The chamber will open automatically.

2

The magazine holds nine 2×1 bricks.

Load the gun by pushing the projectiles under the firing pin.

3

Push down the lid again.

Note how the mechanism locks the lid.

HAPPY SHOOTING!

MINI-THRILLER

★★★★★

SLIDE-ACTION CROSSBOW PISTOL

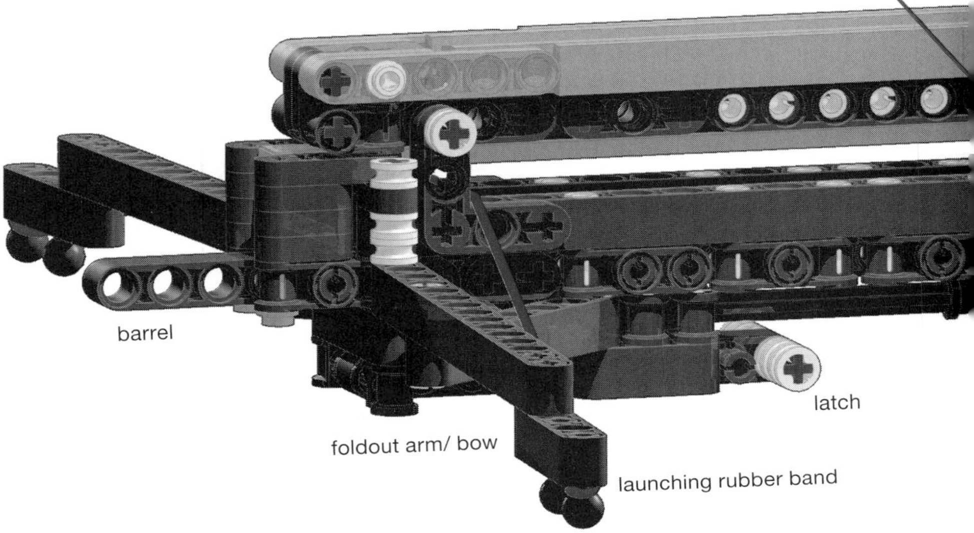

slide

foldout arm/ bow

barrel

foldout arm/ bow

latch

launching rubber band

AMMO: *nine 4×1 bricks*
LENGTH: *30.81 cm*
WIDTH: *29.84/3.98 cm*
HEIGHT: *16.72 cm*
WEIGHT: *275 g*
PARTS: *421*
FIRE POWER: ★★★★★
LEVEL: *Intermediate*

The Mini-Thriller projectile

Retractable bow: The latch at the underside locks the foldout arms in place in both the open and closed positions.

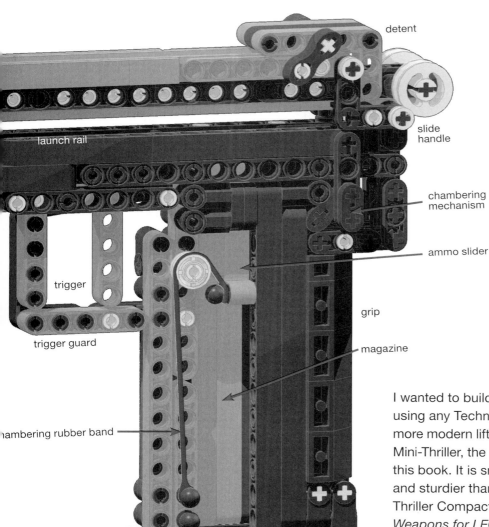

detent

slide
handle

launch rail

chambering
mechanism

ammo slider

trigger

grip

trigger guard

magazine

chambering rubber band

I wanted to build a crossbow-pistol without using any Technic bricks and using only the more modern liftarms. The result was the Mini-Thriller, the last model I designed for this book. It is smaller, more lightweight, and sturdier than its predecessor, the Thriller Compact (from my last book *Weapons for LEGO Lovers*), and even more lightweight than the tiny Liliputt in this book. Because one can build more space-saving and versatile guns with Technic liftarms, I was able to simplify the trigger and chambering mechanisms.

The disadvantage compared to the Thriller Advanced, her older sister, is that the Mini-Thriller has much less punch.

MINI-THRILLER DESIGN HISTORY

The goal of the Mini-Thriller was to expand my LEGO crossbow series with a new model using the newest LEGO Technic parts. The Mini-Thriller was to be smaller, more lightweight, and more reliable than the Thriller Advanced. Also, the Mini-Thriller was to be equipped with a conveniently foldable bow.

Another goal in the design of the Mini-Thriller was to use as few rubber bands as possible, because rubber bands are weak. Mechanically, the Mini-Thriller is a success, but with all its rounded components, it lacks the aggressive look of the Thriller Advanced. The Mini-Thriller also has less fire power.

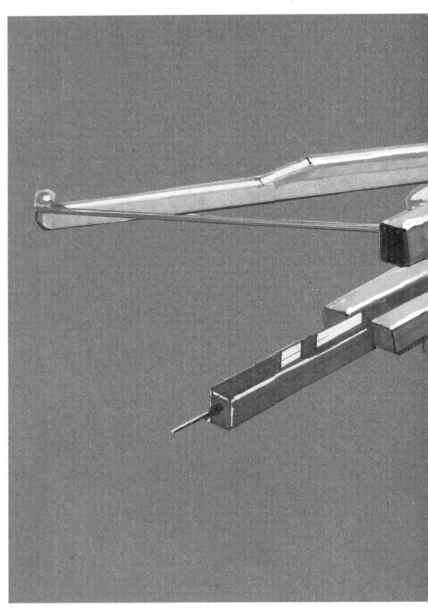

This is the first sketch of the gun. As you can see, the launching bands' mounting arms were first going to attach at the back. I abandoned this idea in the final layout.

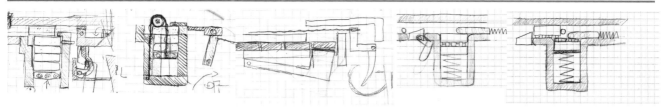

Ideas for new chambering mechanisms

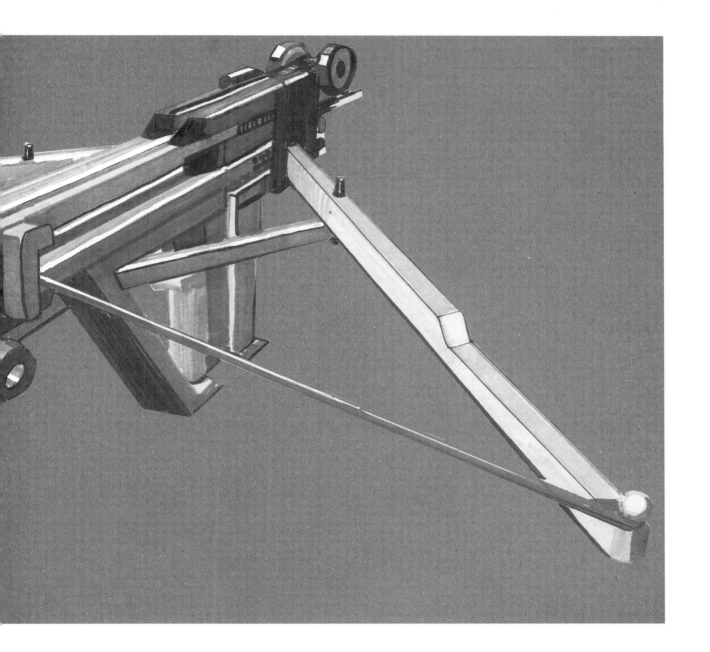

1

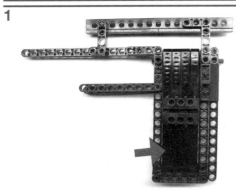

As with the other guns, I designed the magazine first.

2

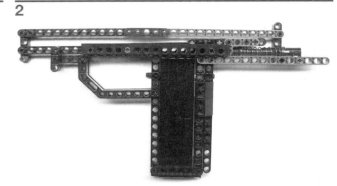

Tests followed, and I added a new chambering mechanism.

3

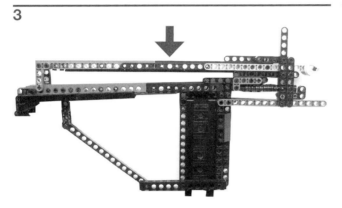

This is the first version with the slide.

4

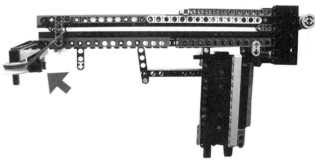

Here is the first version with the bow. I was still using many Technic bricks in this development stage, as you can see.

5

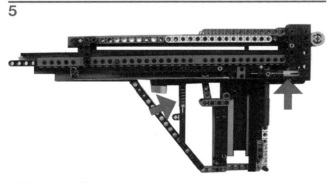

This is the first working version. At this stage, you may notice the return springs used for the trigger and chambering mechanisms.

6

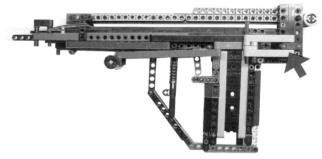

As shown on page 70, I planned to attach the foldable mounting arms for the launching rubber bands at the rear of the weapon. This would have been an interesting variation on the classic T-shaped design of the crossbow. However, because this alternative design was too space intensive, I discarded it.

7

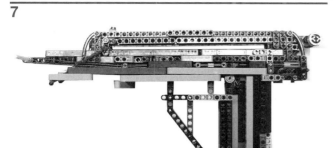

Here I replaced all the Technic bricks with Technic liftarms.

8

From this version on, the Mini-Thriller got a latch to hold the slide in firing position. This device also makes a cool rasping sound when you move the slide. At this point, I still used springs.

9

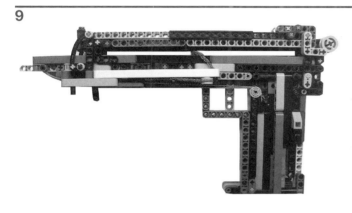

Here I replaced all springs with the rubber part Axle Joiner Double Flexible (45590). You can tell how much space this measure saved by how much shorter the weapon has become behind the handle. I initially planned to use an enlarged trigger guard that completely enclosed the firing hand, but I decided against doing so since this smaller variant seemed handier to me.

10 FINAL VERSION

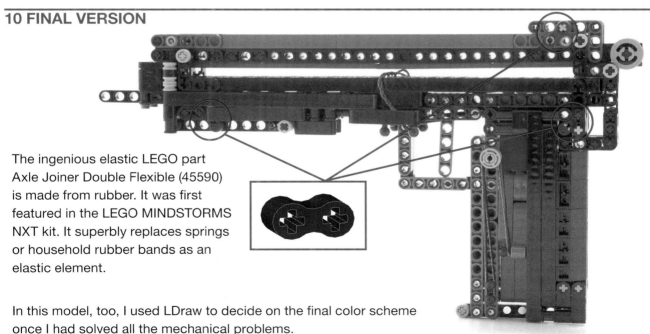

The ingenious elastic LEGO part Axle Joiner Double Flexible (45590) is made from rubber. It was first featured in the LEGO MINDSTORMS NXT kit. It superbly replaces springs or household rubber bands as an elastic element.

In this model, too, I used LDraw to decide on the final color scheme once I had solved all the mechanical problems.

HOW IT WORKS

1

Initial position: The L-shaped lever blocks the projectile from gliding up into the path of the slide, so that it can transport the launching rubber band to behind the magazine.

The bricks in the magazine alternate studs-down and studs-up to make sure that they can't connect to each other.

2

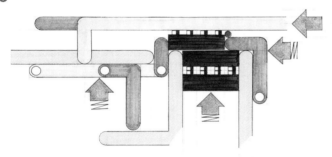

By pulling back the slide, the L-shaped lever is pushed backwards, so that the topmost projectile can slip into the launch rail.

3

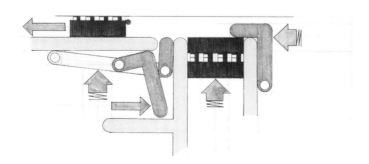

The launching rubber band hooks behind the top projectile, held in place by the trigger mechanism. The Mini-Thriller is now ready to be fired.

4

Pulling the trigger clears the launch rail. The launching rubber band hurls the projectile forward.

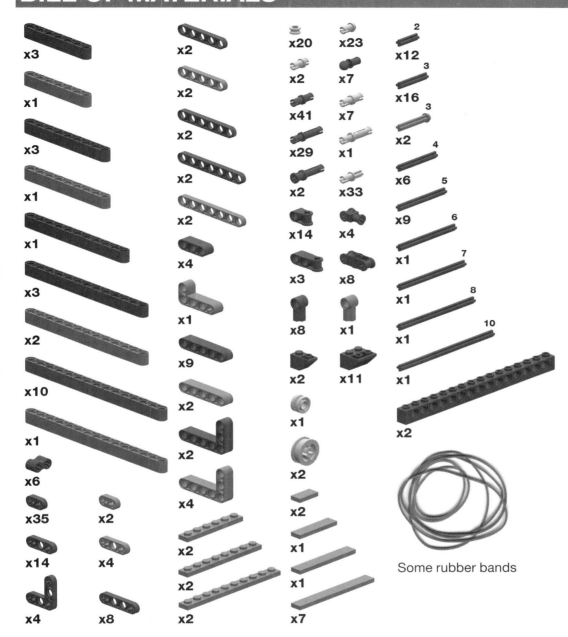

BILL OF MATERIALS

Some rubber bands

MAGAZINE

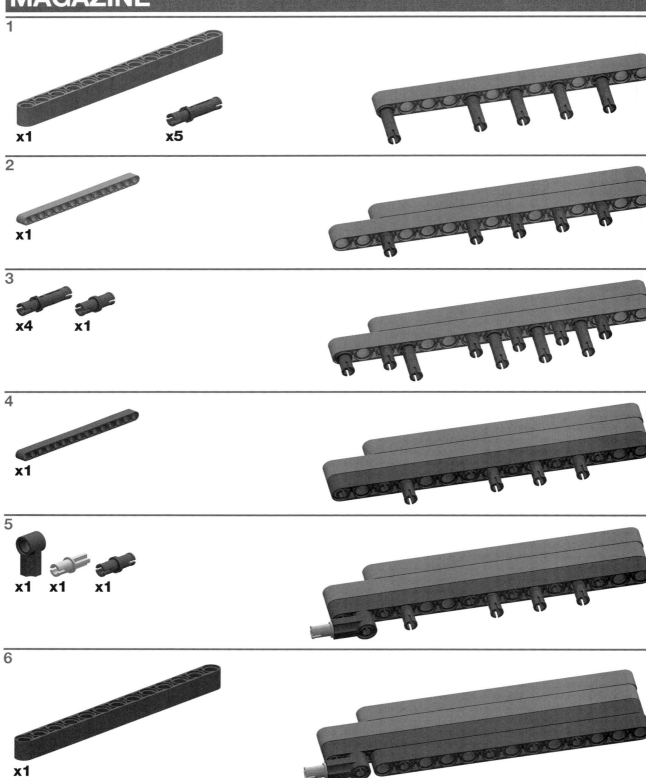

1
x1   x5

2
x1

3
x4   x1

4
x1

5
x1   x1   x1

6
x1

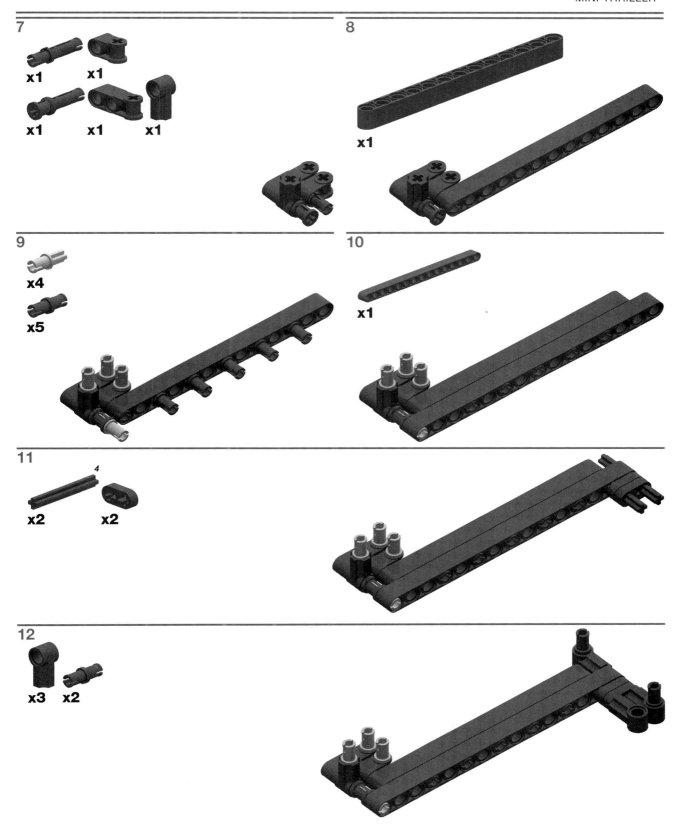

7

x1 x1

x1 x1 x1

8

x1

9

x4

x5

10

x1

11

x2 x2

12

x3 x2

13

x1

x1

x1

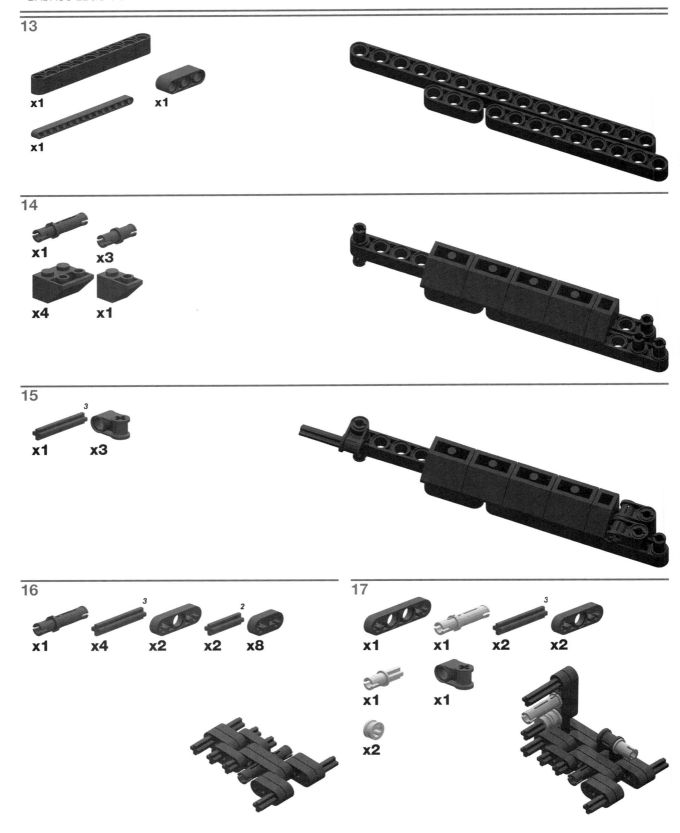

14

x1

x3

x4

x1

15

3

x1

x3

16

x1

x4

3

x2

2

x2

x8

17

x1

x1

3

x2

x2

x1

x1

x2

18

x1

x1

x1

19

x1 x7

x3

20

x3

x3

21

7

x1

6

x1

5

x2

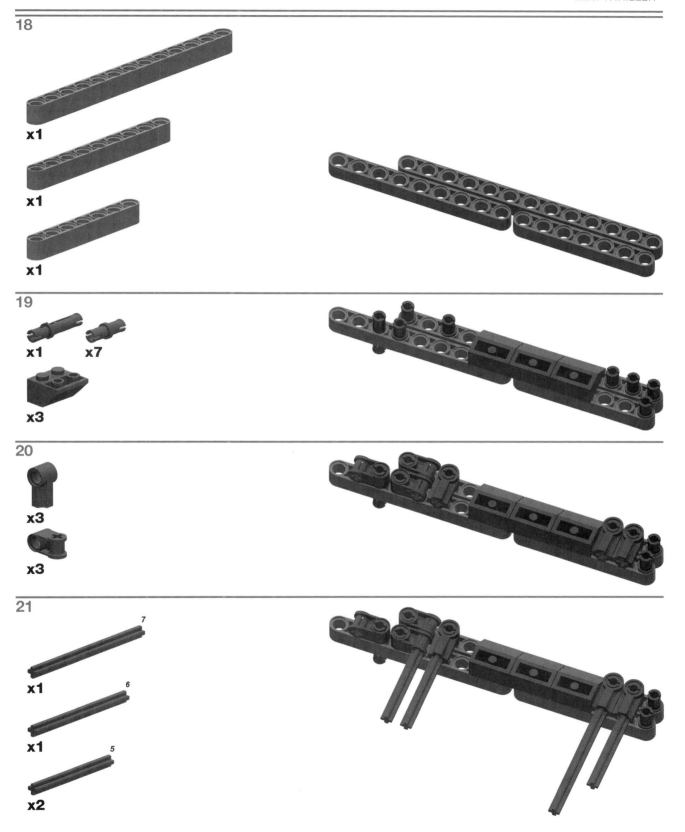

22

x1

23

x1

24

x1

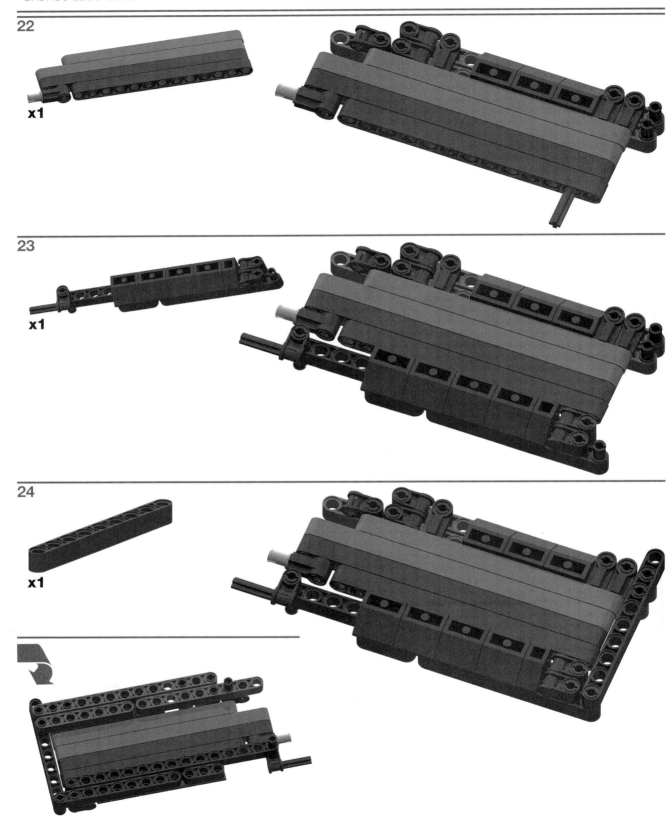

25

x2 x1

x1 x1

x2

26

x1

27

x4 x1

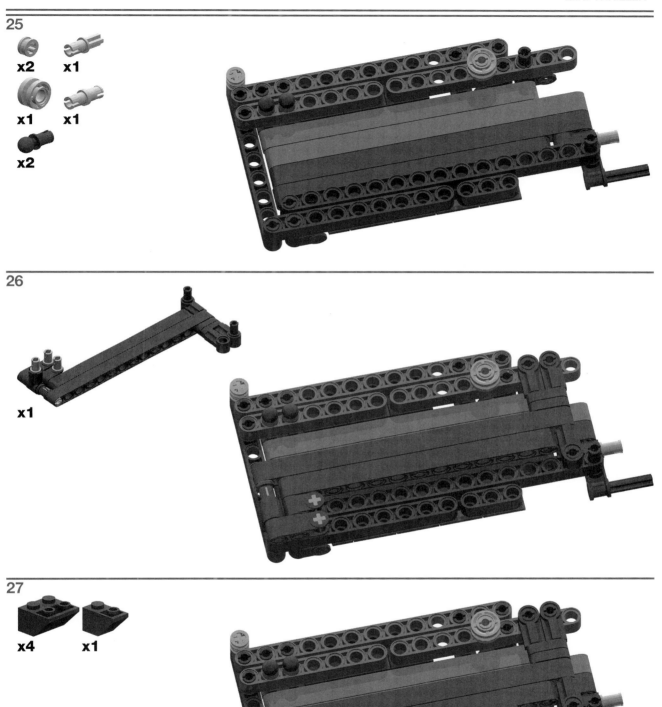

28

x1

29

x1

30

5

3

x1   **x1**

31

x1

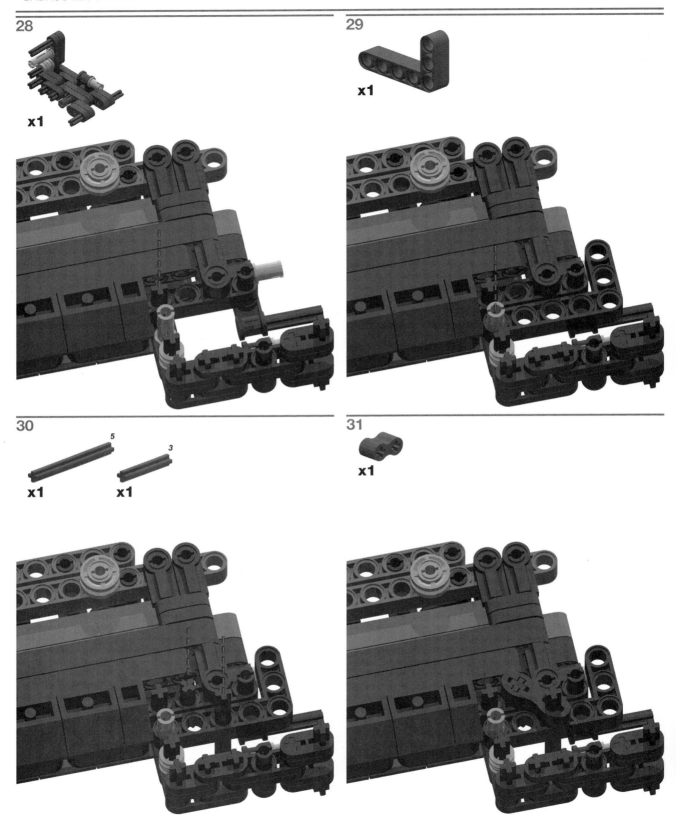

32

x1   x2

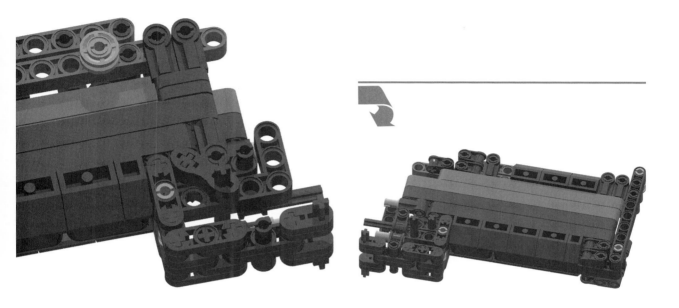

33

x1

34

5

x1

35

x1   x1

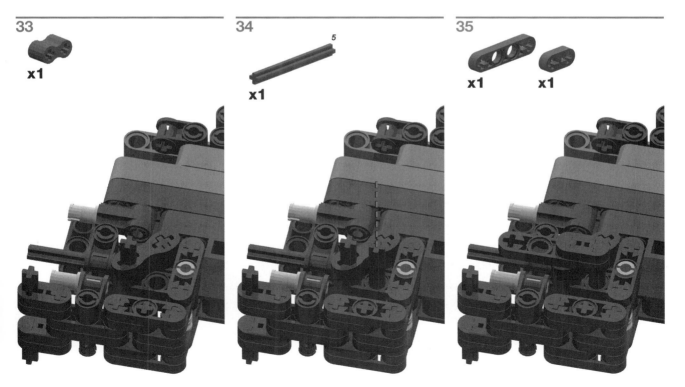

SLIDE

1

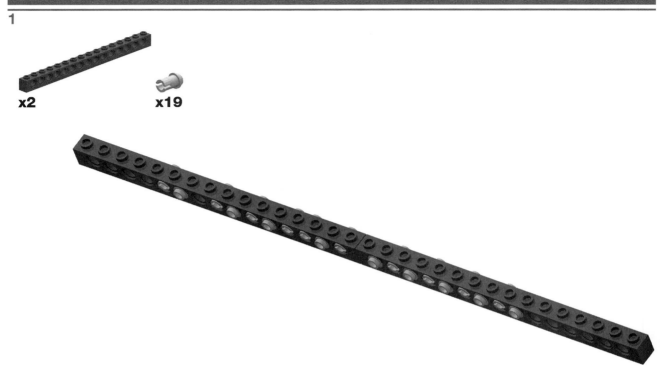

x2   x19

2

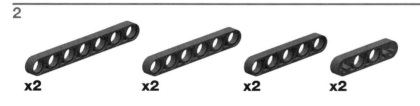

x2   x2   x2   x2

3

x1 x1 x2 x6 x1

4

x2 x1

5

x2   x2   x2   x2

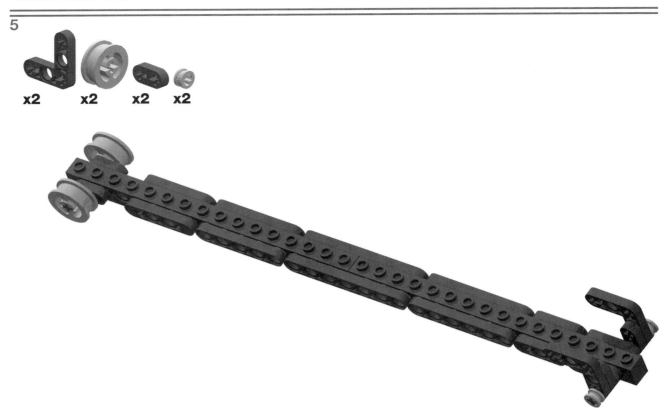

6

x3   x1   x1

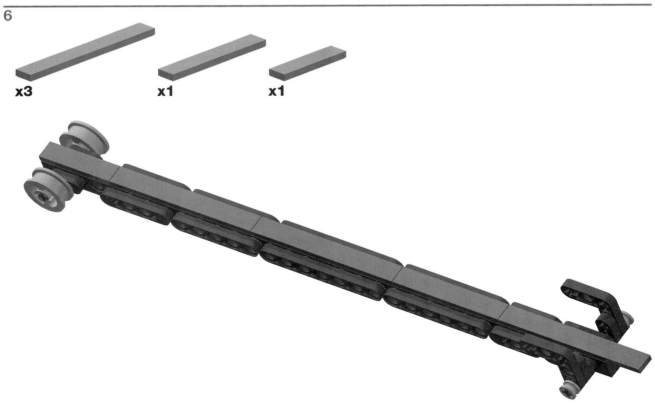

UPPER SECTION

1

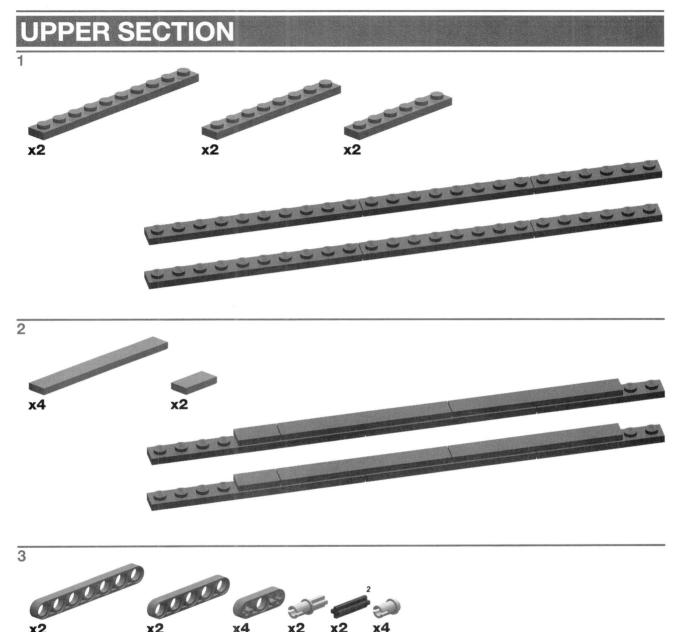

x2

x2

x2

2

x4

x2

3

x2

x2

x4

x2

x2

2

x4

4

x1   *3* x2

5

x2

x1

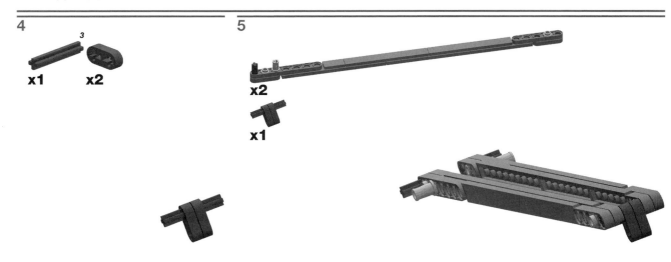

LAUNCH RAIL

1

x1

x1

3 x2

x7

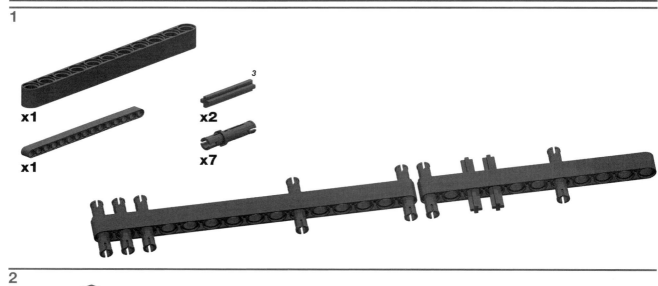

2

x1   *3* x2   x2

x1   x4   x5

3

x2 x1 x2 x1 x6

4

x4 x2 x8 x1

5

x2 x1 x1

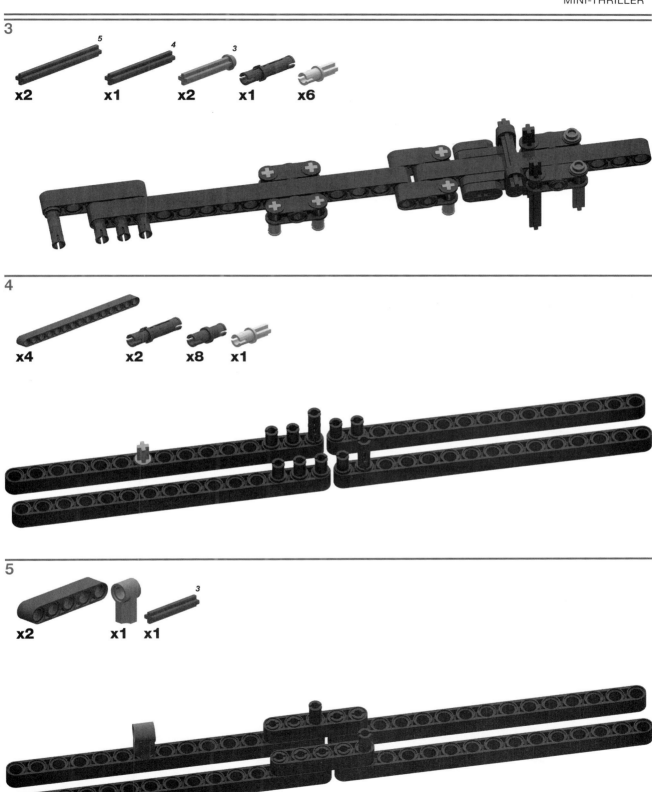

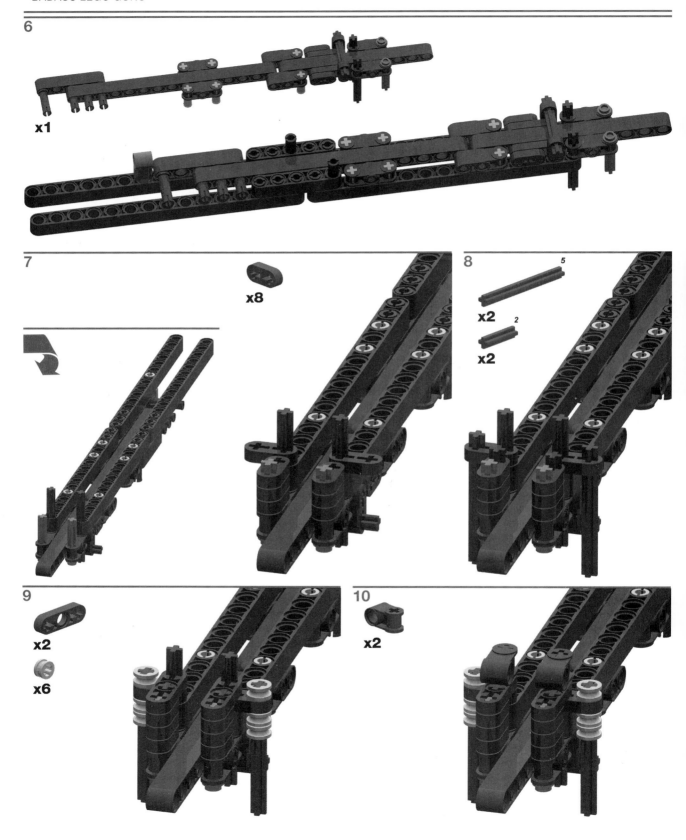

6

x1

7

x8

8

5

x2

2

x2

9

x2

x6

10

x2

FINAL ASSEMBLY

1

x2

x2

x2

x4

x2

2

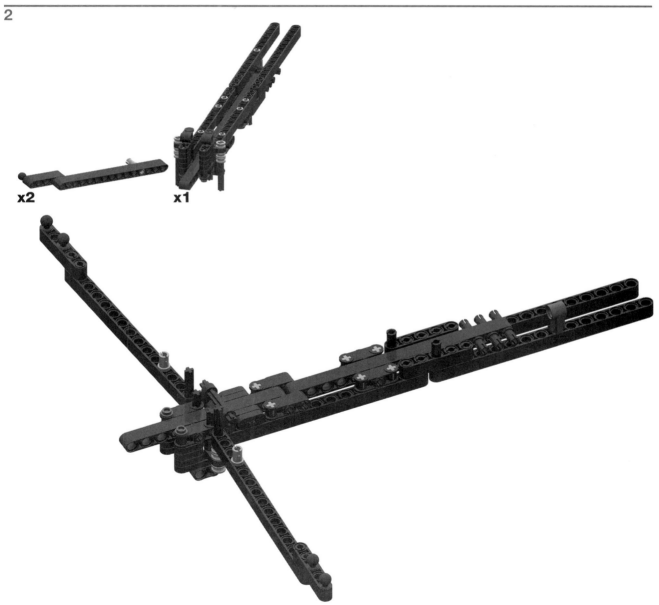

x2

x1

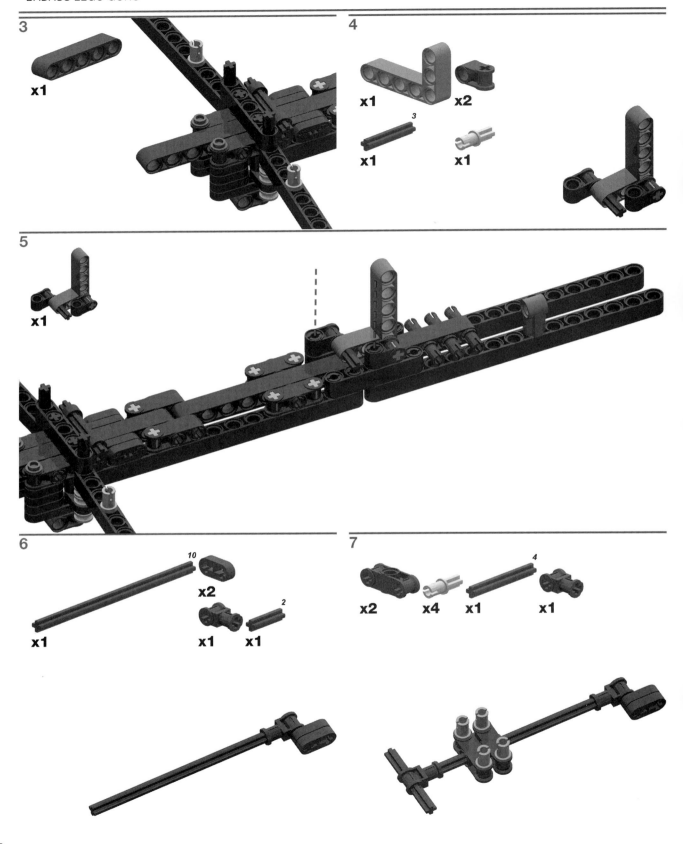

3

x1

4

x1   x2

3

x1   x1

5

x1

6

10

x2

x1   x1   x1

7

4

x2   x4   x1   x1

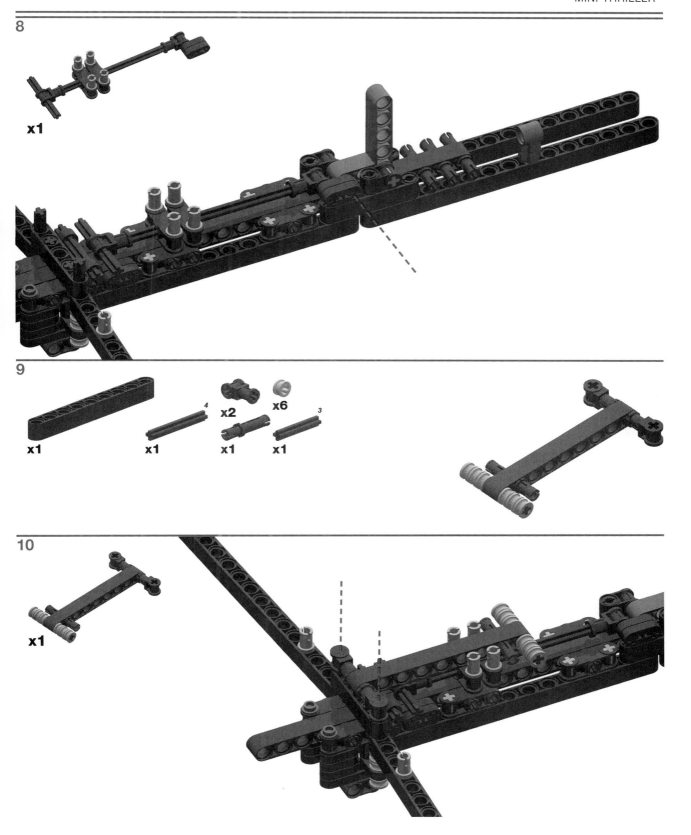

8

x1

9

x1

4

x1

x2

x1

x6

3

x1

10

x1

11

x1 3

12

x2

13

x2

14

x2 **x2**

15

x2

16

x1

x2

x1

x1

x1

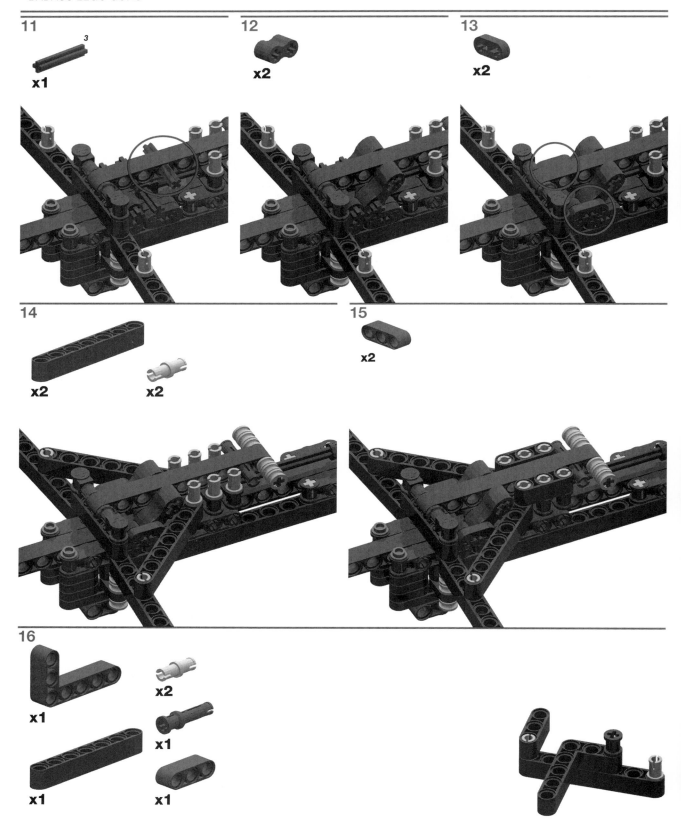

17

x1

18

x1

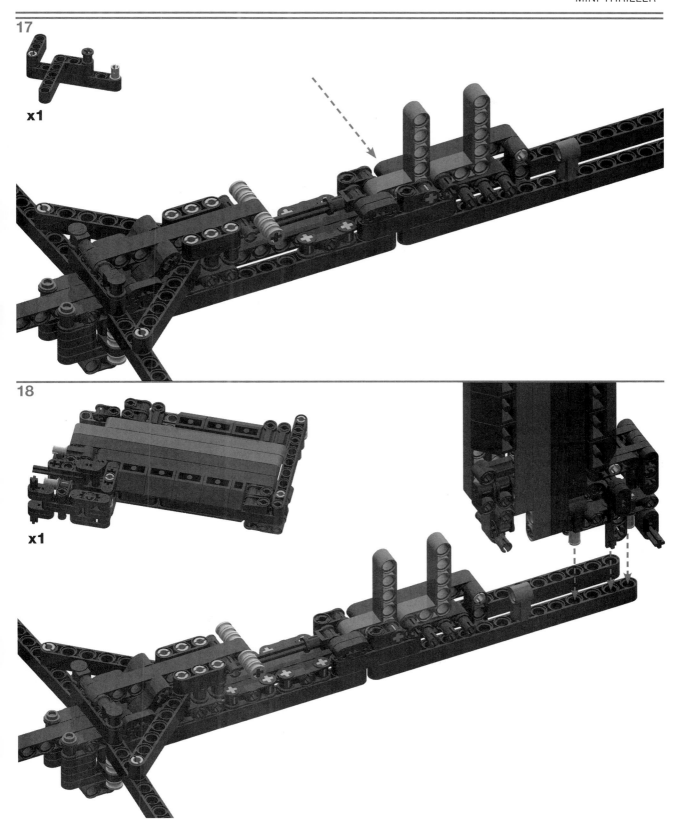

19

x1

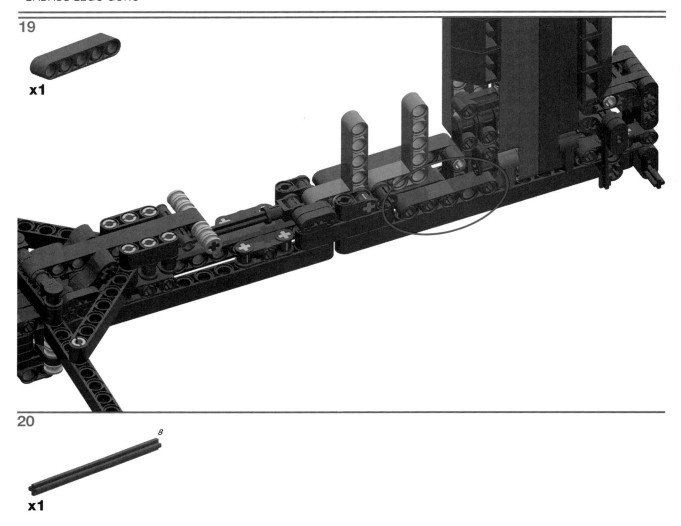

20

8

x1

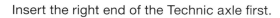

Insert the right end of the Technic axle first.

Push the bottom upward and let it snap onto the trigger component. The axle will now serve as a retaining spring.

21

x1 x2 x9 x2

22

x1

23

x1

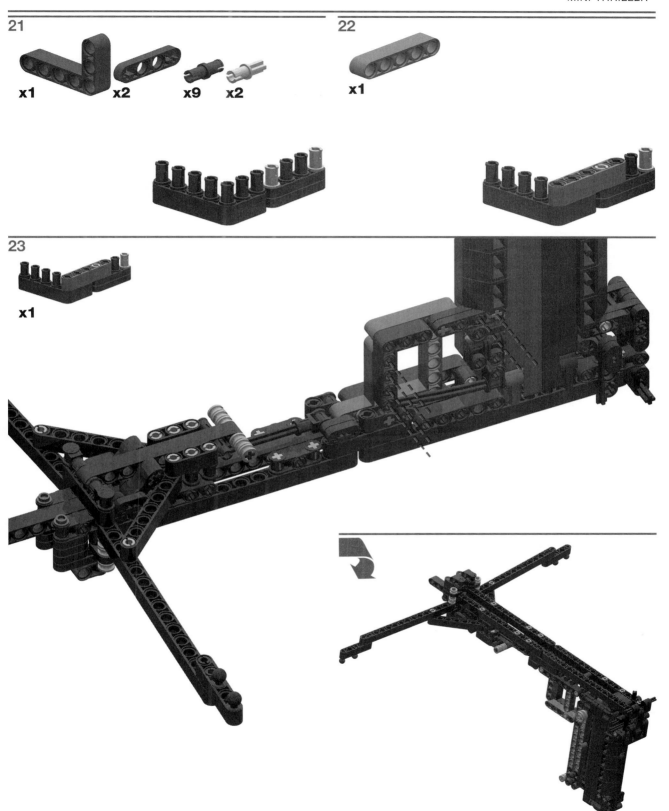

24

x1

25

x1   **x1**

26

x2

x2

x1

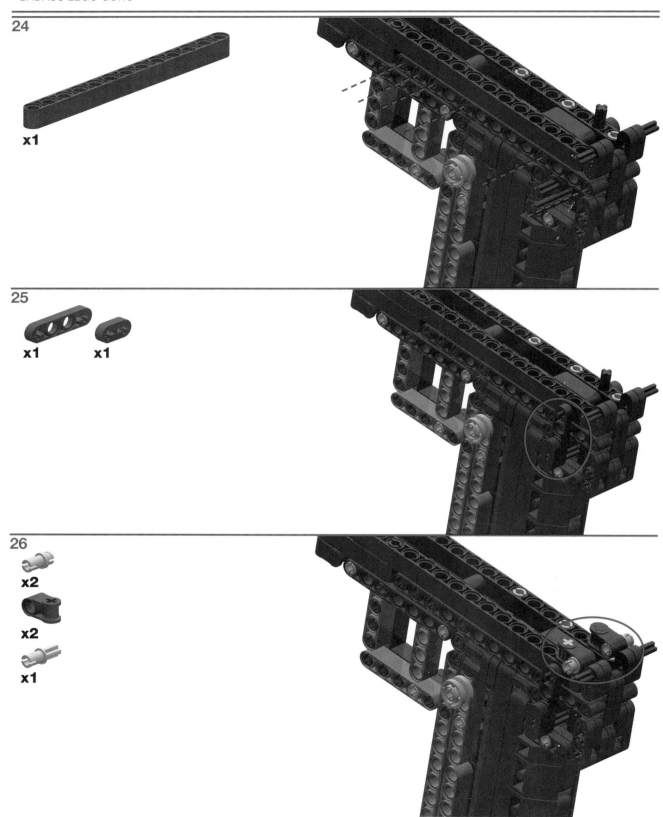

27

x1

28

x1

3

x1

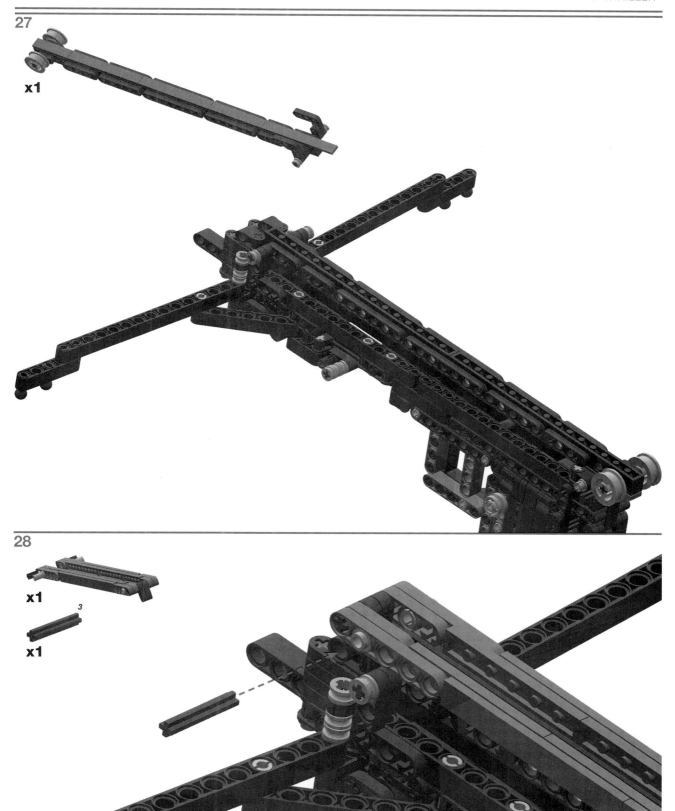

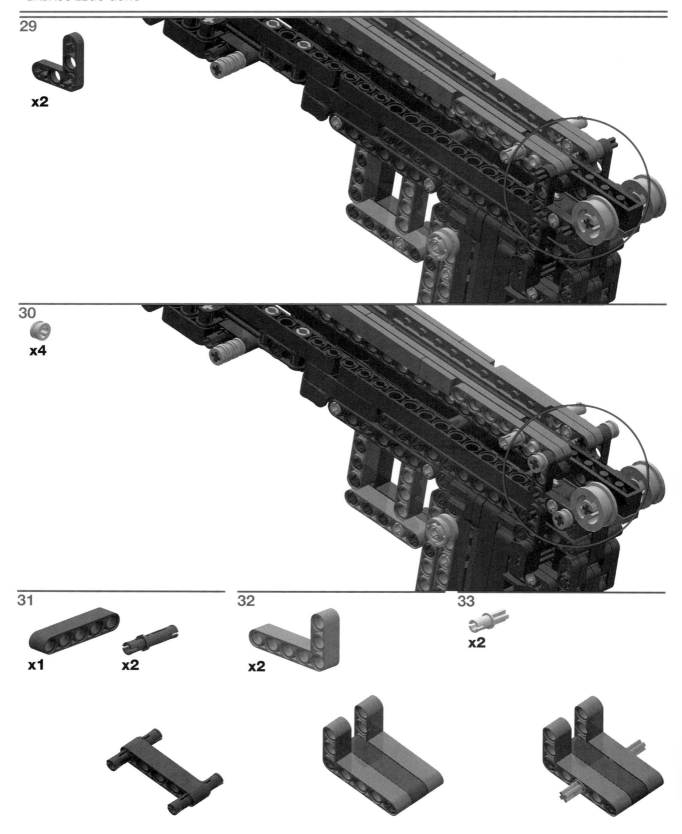

29

x2

30

x4

31

x1   x2

32

x2

33

x2

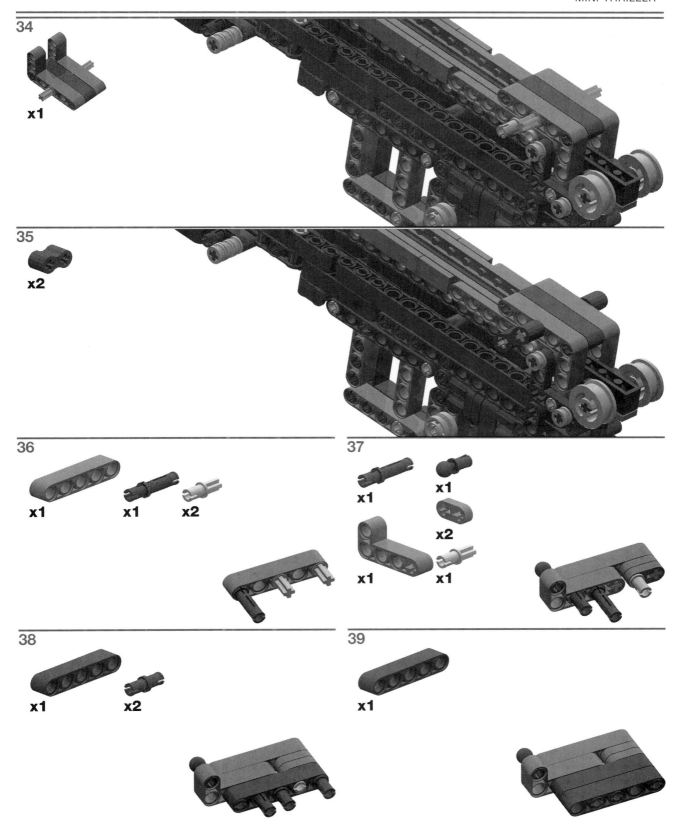

34

x1

35

x2

36

x1   x1   x2

37

x1   x1

x2

x1   x1

38

x1   x2

39

x1

40

x1

x1

LOADING THE GUN

Load the 4×1 bricks so that they alternate: studs on top, stud on bottom, studs on top, studs on bottom, and so on. This prevents them from snapping together.

ATTACHING THE LAUNCHING BAND

1

Attach the rubber band to the right pin, and then pull it through the launch rail.

2

Now twist it. That's important for the trigger mechanism to work well.

3

Attach the other end of the rubber band to the left pin, and your Mini-Thriller is ready to fire!

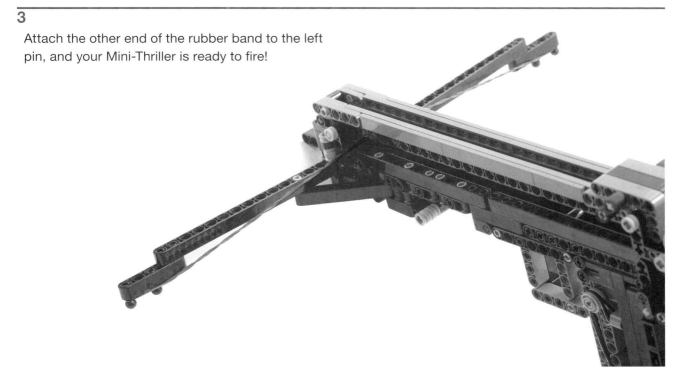

THRILLER

★★★★★

SLIDE-ACTION CROSSBOW PISTOL

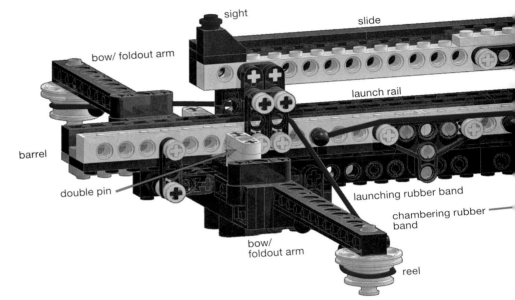

AMMO: *nine bricks*
LENGTH: *39.77 cm*
WIDTH: *34.13/8.47 cm*
HEIGHT: *19.73 cm*
WEIGHT: *550 g*
PARTS: *582*
FIRE POWER: ★★★★★
LEVEL: *Expert*

The Thriller Advanced projectile: a 4×1 brick, a plate, and a tile

The retractable bow: With the two double-pins you can lock the two foldout arms of the bow in the folded-out or retracted position.

ADVANCED

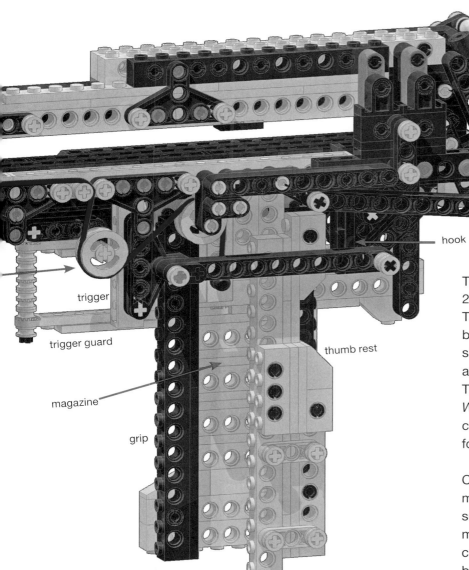

latch

rocker

slide handle

chambering
mechanism

hook

trigger

trigger guard

magazine

grip

thumb rest

The Thriller Advanced has a 20-year development history. The original model came into being in 1990, followed by versions with a magazine (1994) and a slide (1997). In 2007, the Thriller Automatic was built for *Weapons for LEGO Lovers*. It can be seen as the prototype for the Thriller Advanced.

Considering this long development history, it should hardly be surprising that this model is the most complex and technologically sophisticated one in this book. This model is the most precise, has the longest range, and, thanks to its projectile's large caliber, packs the most powerful punch of them all.

NOTE: *Building this model requires gluing and sanding a few LEGO pieces.*

THRILLER ADVANCED DESIGN HISTORY

The goal of the Thriller Advanced was to build an improved version of my old Thriller Automatic using modern LEGO Technic liftarms and the like.

1

Thriller Automatic, as seen in my YouTube video, *Slide Action LEGO Crossbow Pistol*

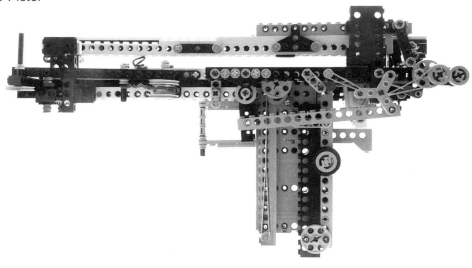

2

The improved version has modified trigger and chambering mechanisms.

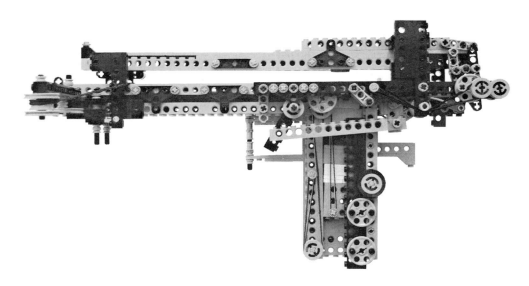

3

Thriller Automatic from my book *Weapons for LEGO Lovers*

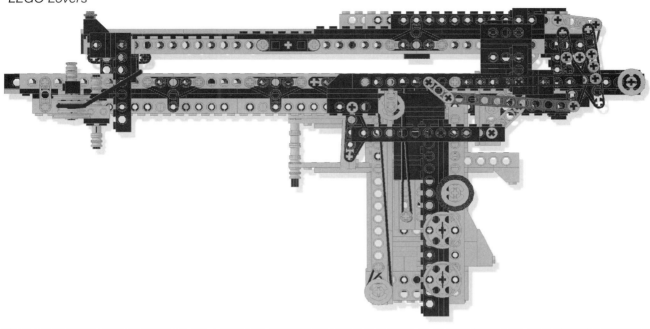

4

The final version is lightweight and more reliable. Additionally, the handle was moved backward, increasing the length of the barrel.

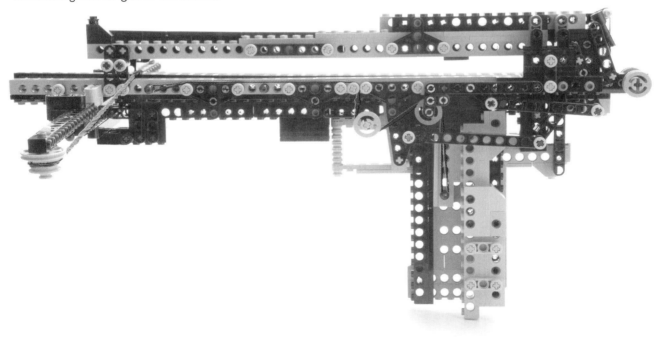

HOW IT WORKS

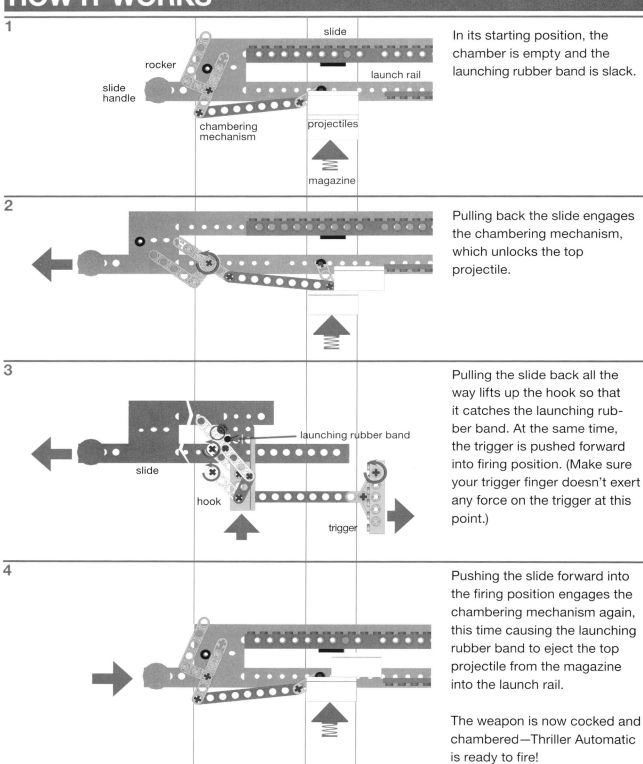

1 In its starting position, the chamber is empty and the launching rubber band is slack.

2 Pulling back the slide engages the chambering mechanism, which unlocks the top projectile.

3 Pulling the slide back all the way lifts up the hook so that it catches the launching rubber band. At the same time, the trigger is pushed forward into firing position. (Make sure your trigger finger doesn't exert any force on the trigger at this point.)

4 Pushing the slide forward into the firing position engages the chambering mechanism again, this time causing the launching rubber band to eject the top projectile from the magazine into the launch rail.

The weapon is now cocked and chambered—Thriller Automatic is ready to fire!

BILL OF MATERIALS

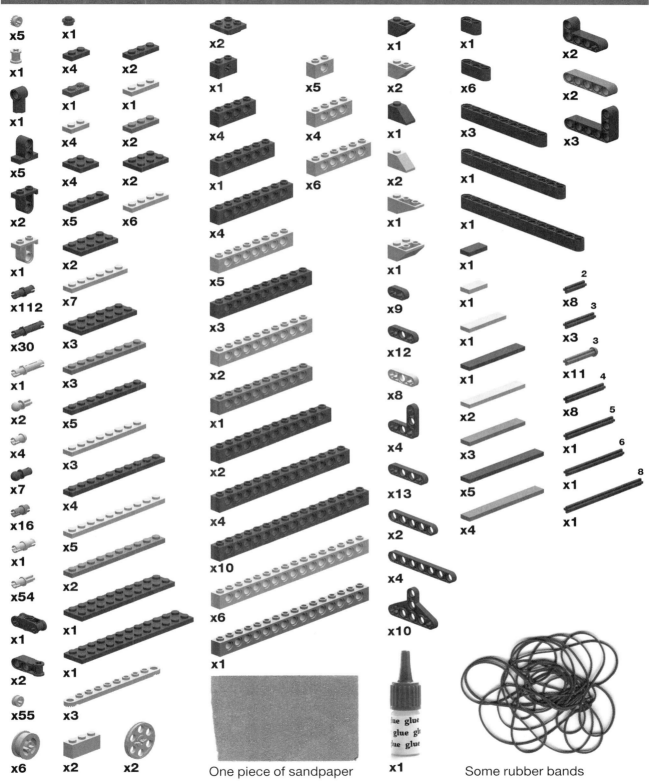

x5
x1
x1
x1
x5
x2
x1
x112
x30
x1
x2
x4
x7
x16
x1
x54
x1
x2
x55
x6

x1
x4
x1
x4
x4
x5
x2
x7
x3
x3
x5
x3
x4
x5
x2
x1
x1
x3
x2
x2

x2
x1
x4
x2
x6

x2
x1
x5
x4
x4
x5
x3
x2
x1
x2
x4
x10
x6
x1

x5
x4
x6

x1
x2
x1
x2
x1
x1
x1
x9
x12
x8
x4
x13
x2
x4
x10

x1
x6
x3
x1
x1
x1
x1
x1
x2
x3
x5
x4

x2
x2
x3

2
x8
3
x3 3
3
x11
4
x8
5
x1
6
x1
8
x1

One piece of sandpaper

glue glue
glue gl
glue glue
lue glue
x1

Some rubber bands

109

MAGAZINE

1

x3

x3

x6

x2

x3

x3

x4

x1

x2

x1

x1

x1

x2

x2

x1

x1

x1

x12

x1

5

x3

x5

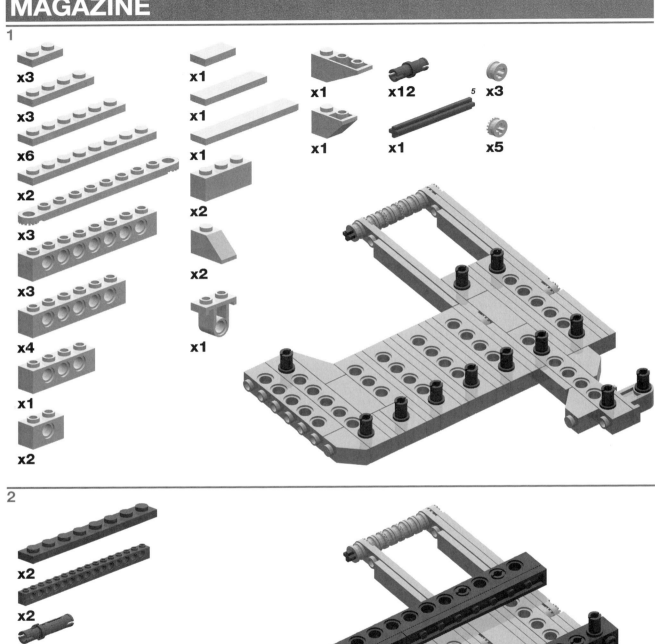

2

x2

x2

x1

x2

3

x1

x1

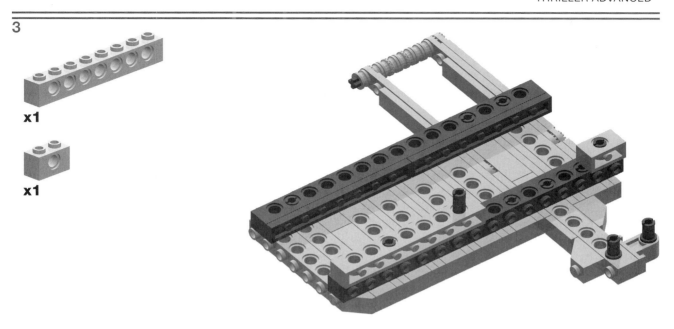

4

x2

x1

x1

x1

x1

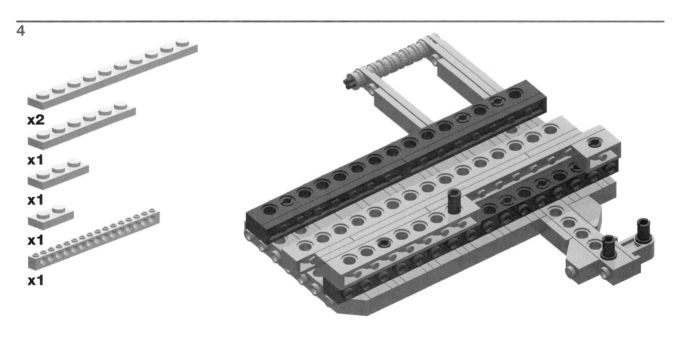

5

x4

x3

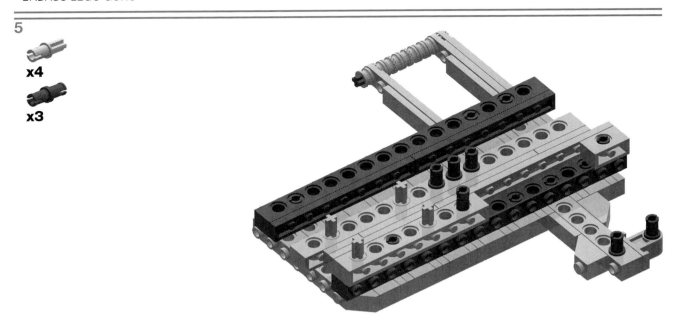

6

x1

x2

x1

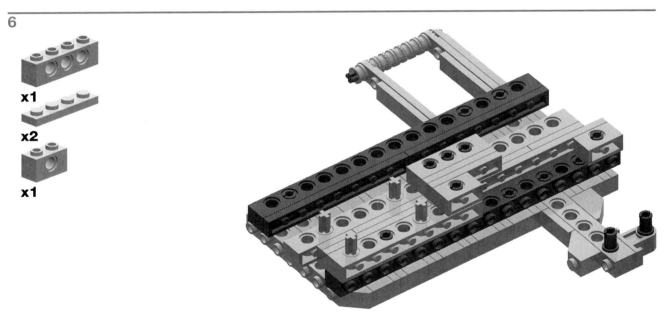

7

x2

x4

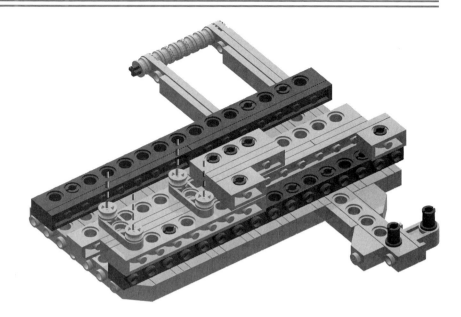

8

x1

x1

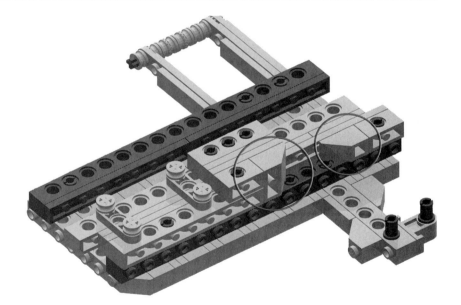

LAUNCH RAIL: UPPER SECTION

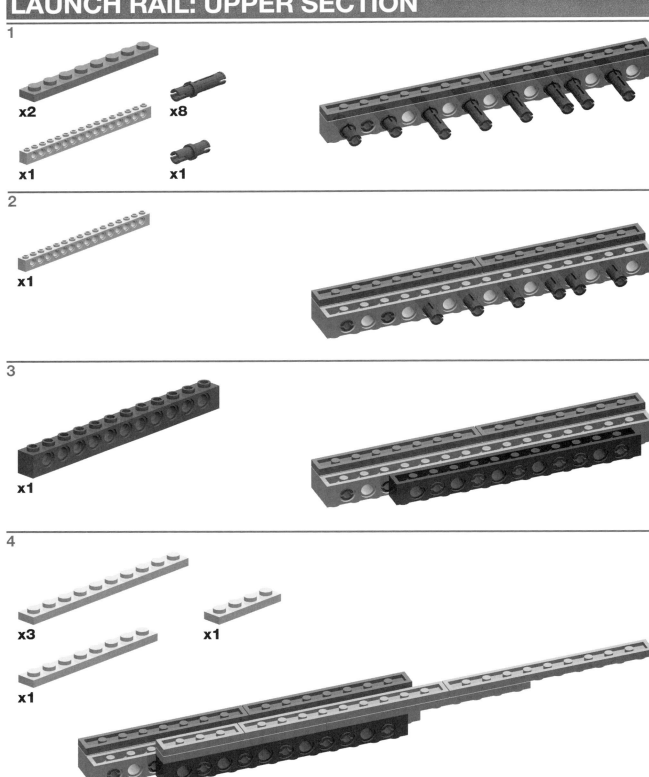

1

x2

x8

x1

x1

2

x1

3

x1

4

x3

x1

x1

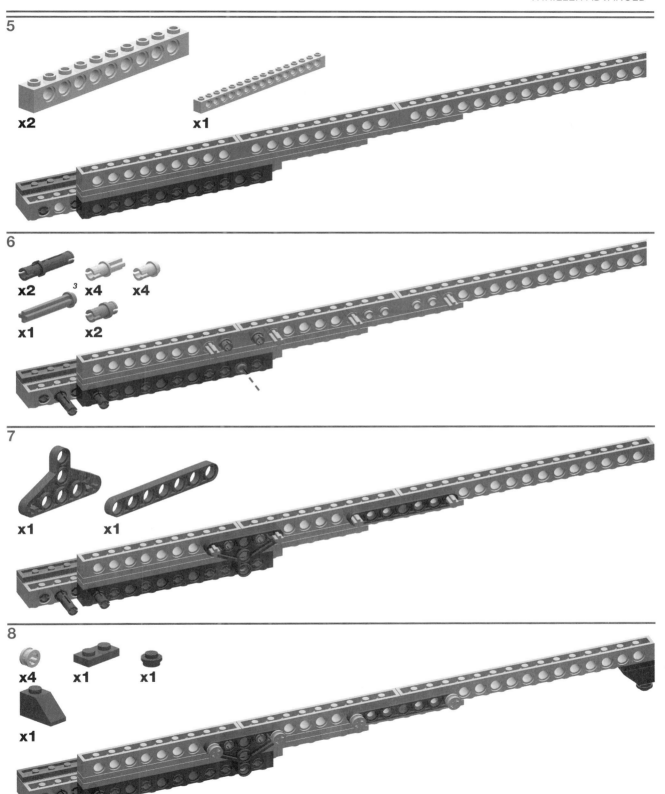

5

x2   x1

6

x2   3 x4   x4
x1   x2

7

x1   x1

8

x4   x1   x1
x1

LEFT SECTION

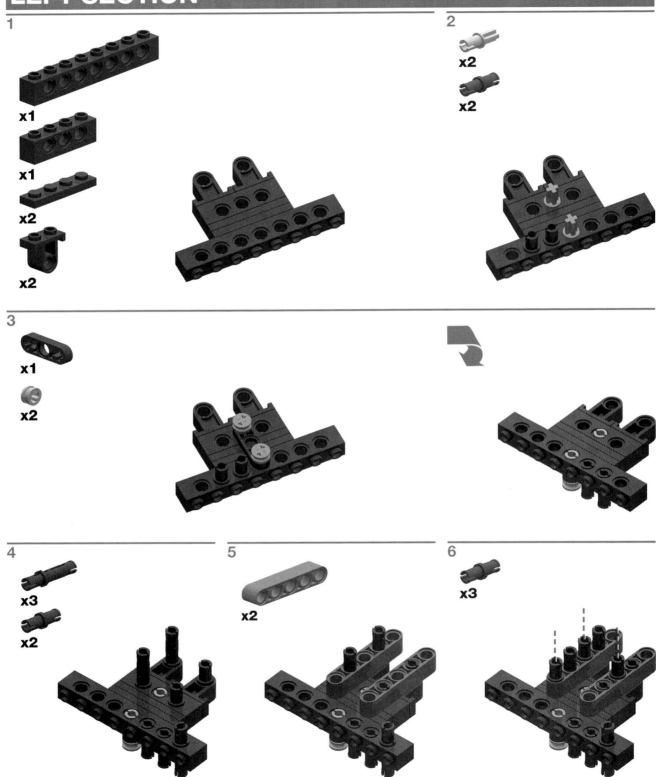

LEFT FOLDOUT ARM

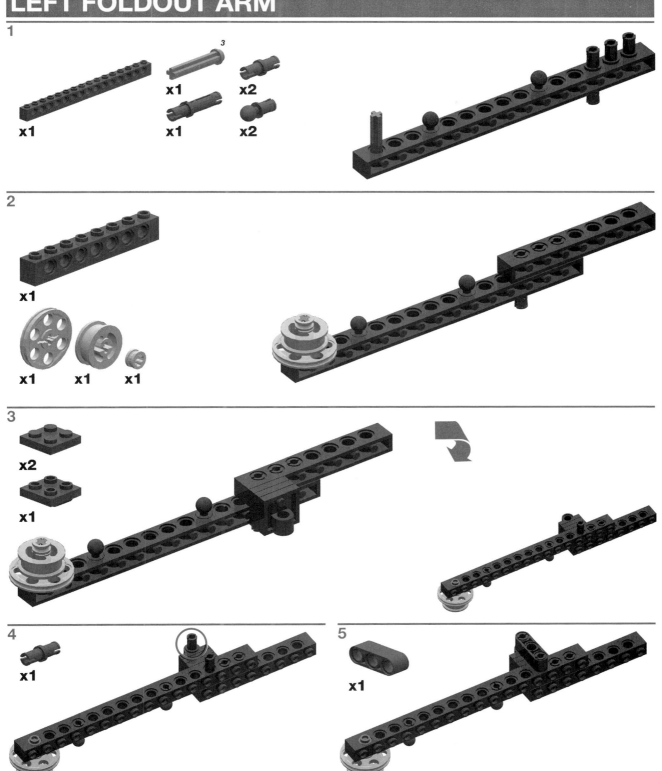

RIGHT FOLDOUT ARM

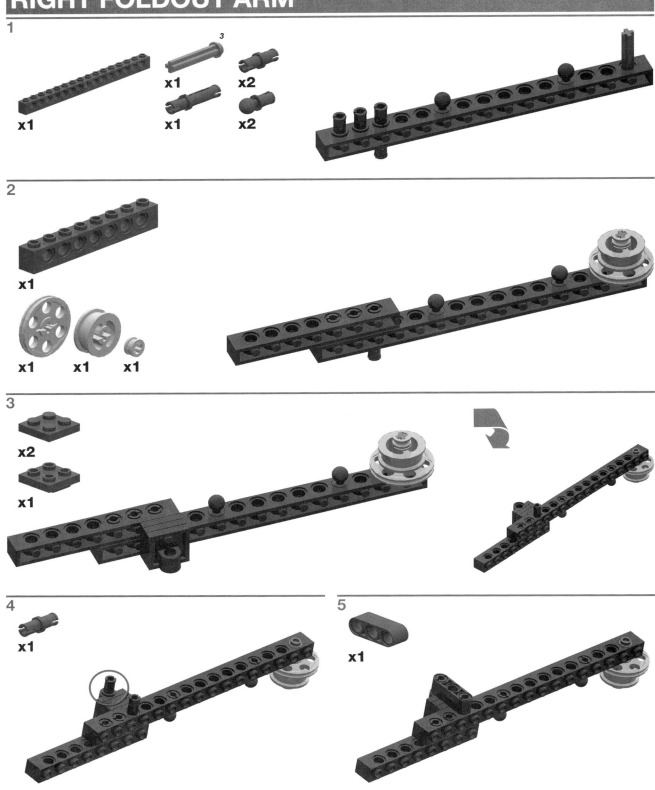

SLIDE

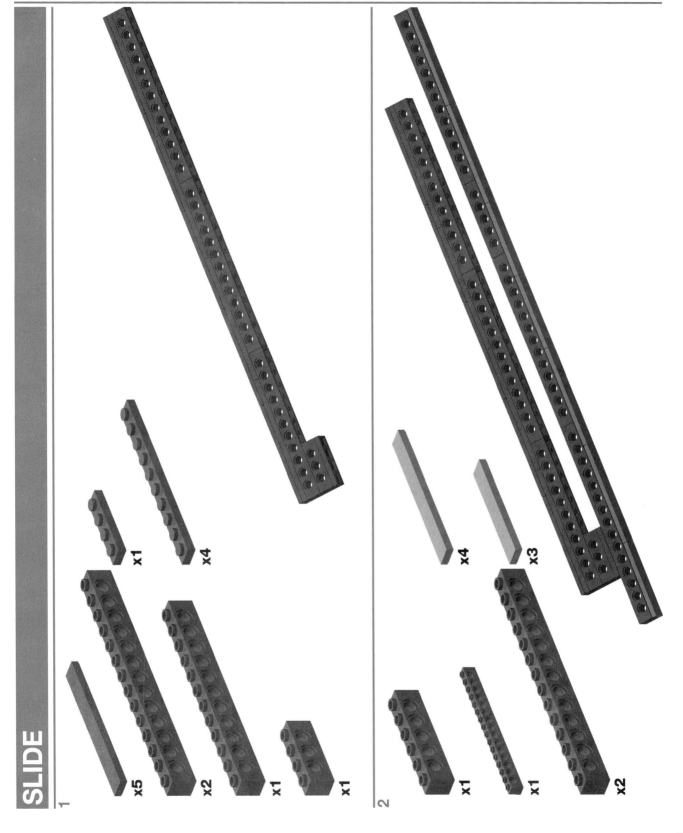

1

x5
x2
x1
x1
x1
x4

2

x1
x1
x2
x4
x3

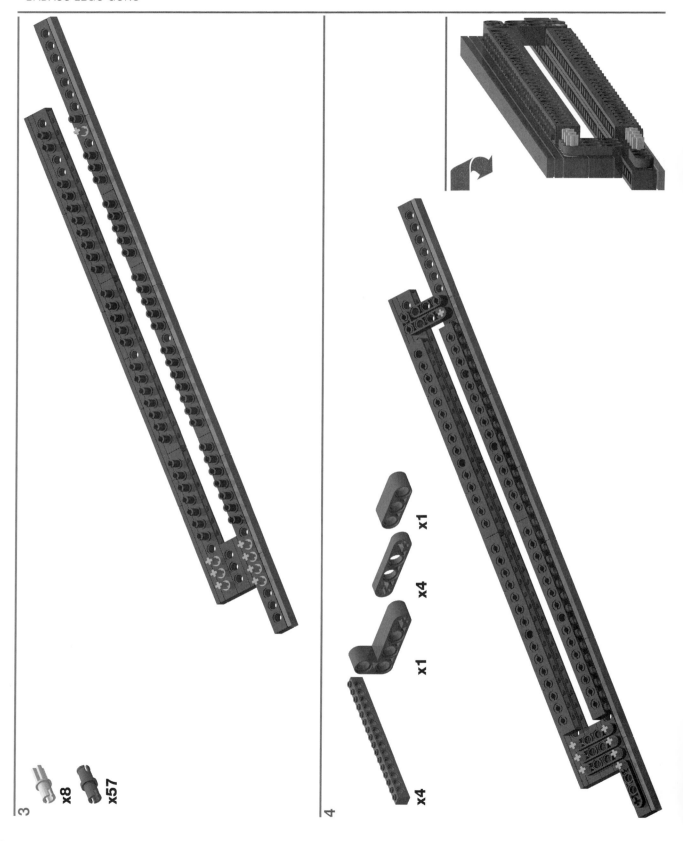

3

x8

x57

4

x1

x4

x1

x4

FRONT SHAFT

1

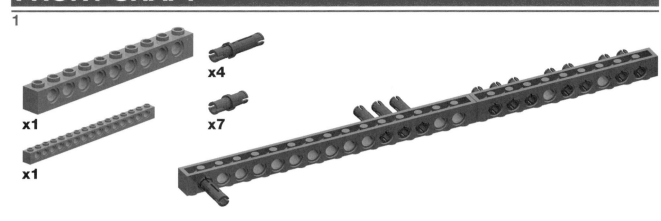

x1

x1

x4

x7

2

x3

3

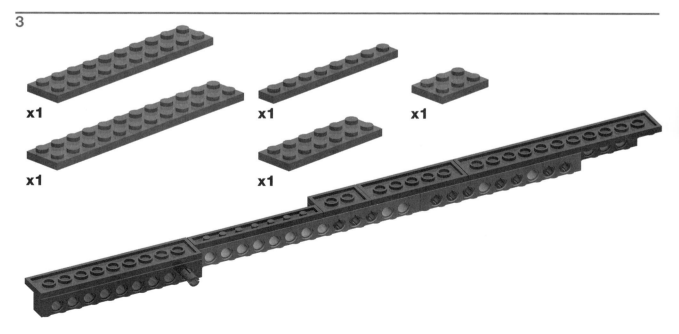

x1

x1

x1

x1

x1

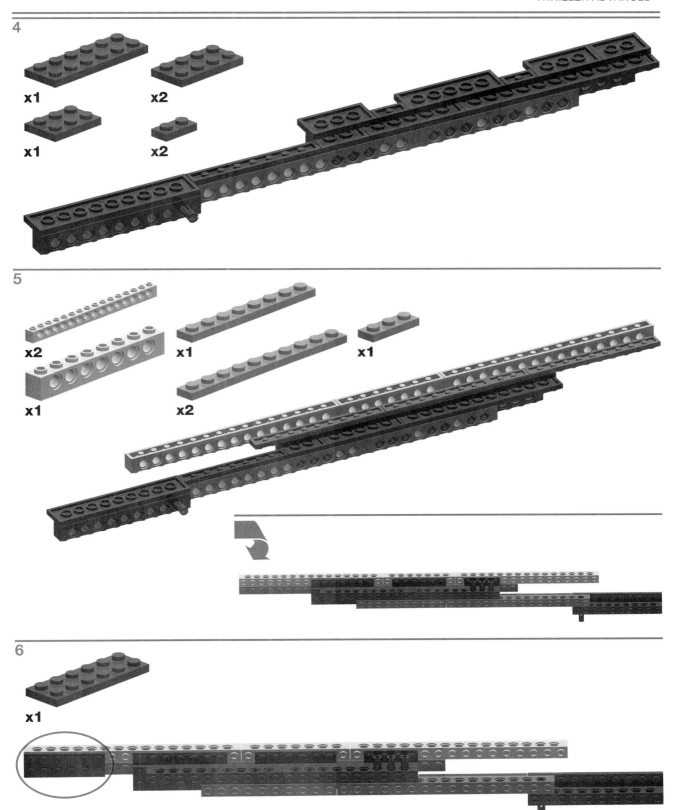

4

x1

x2

x1

x2

5

x2

x1

x1

x1

x2

6

x1

7

x2 x1 x1

x1 x1

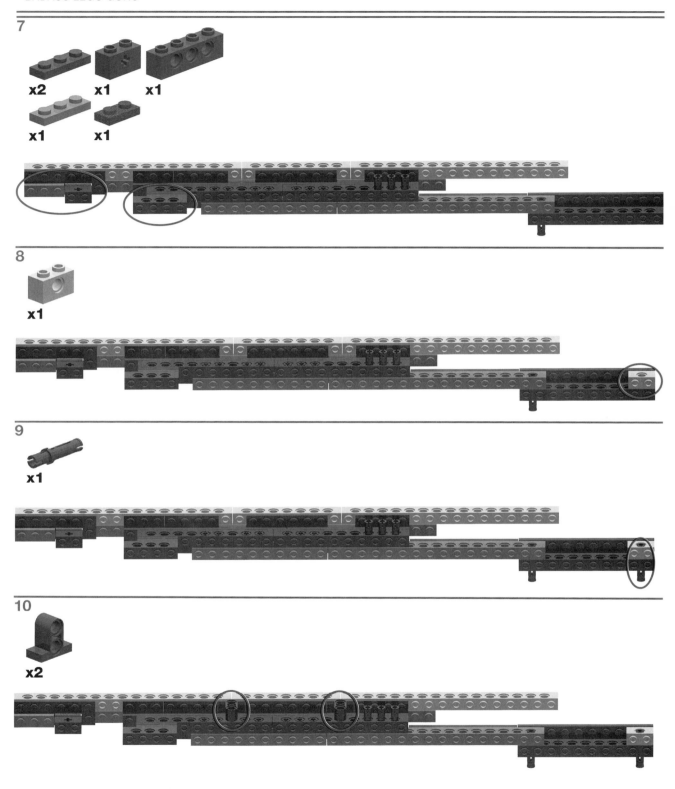

8

x1

9

x1

10

x2

11

x1 ₃x9 x5

x2 x2

12

x2 x1 x4

13

x1 x1

x1 x4

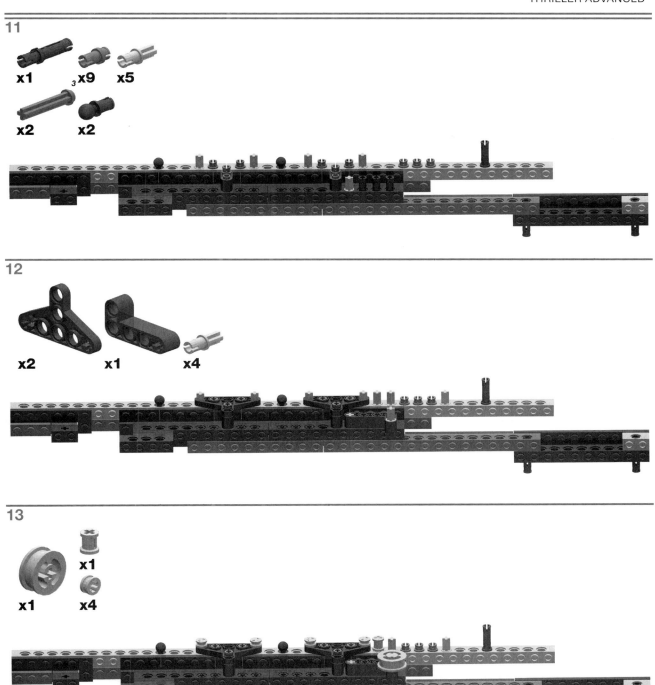

14

x2

x3

2

4

x1

x1

x1

15

x1

x1

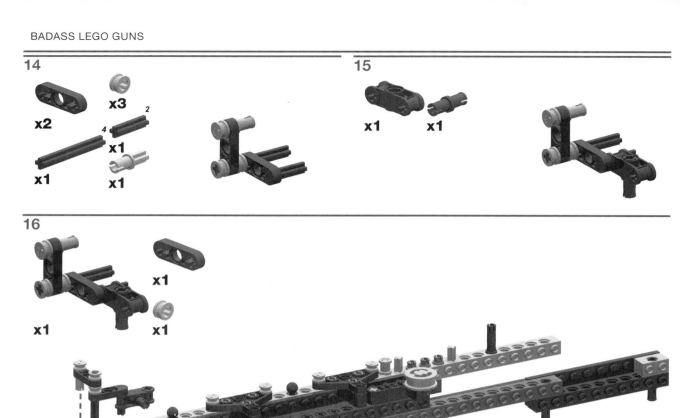

16

x1

x1

x1

x1

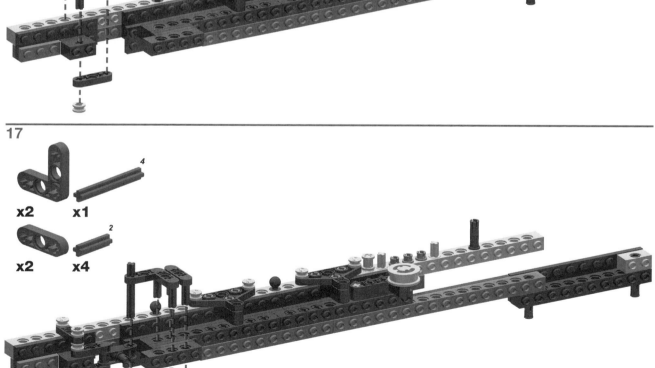

17

x2

x1

4

x2

x4

2

FINAL ASSEMBLY

1

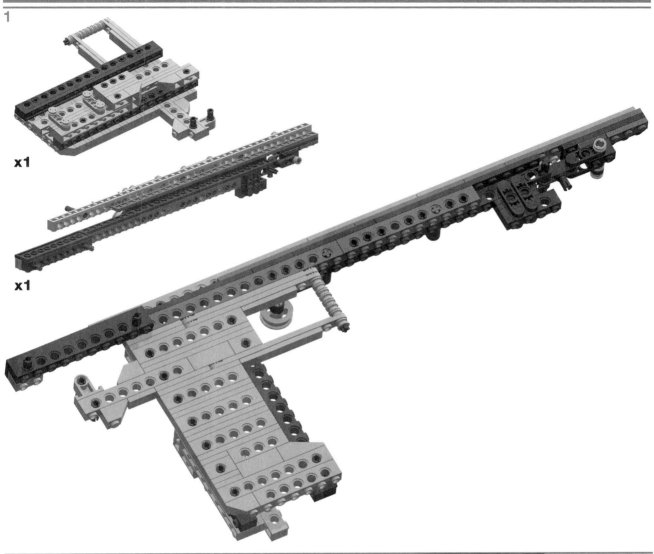

x1

x1

2

x5   x2   x2   x1

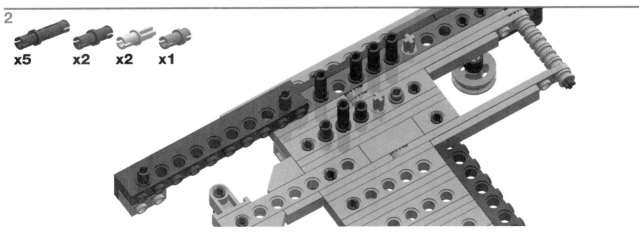

3

x1

x3

4

x1

x1

5

x2

x2

x3

x3

6

x1

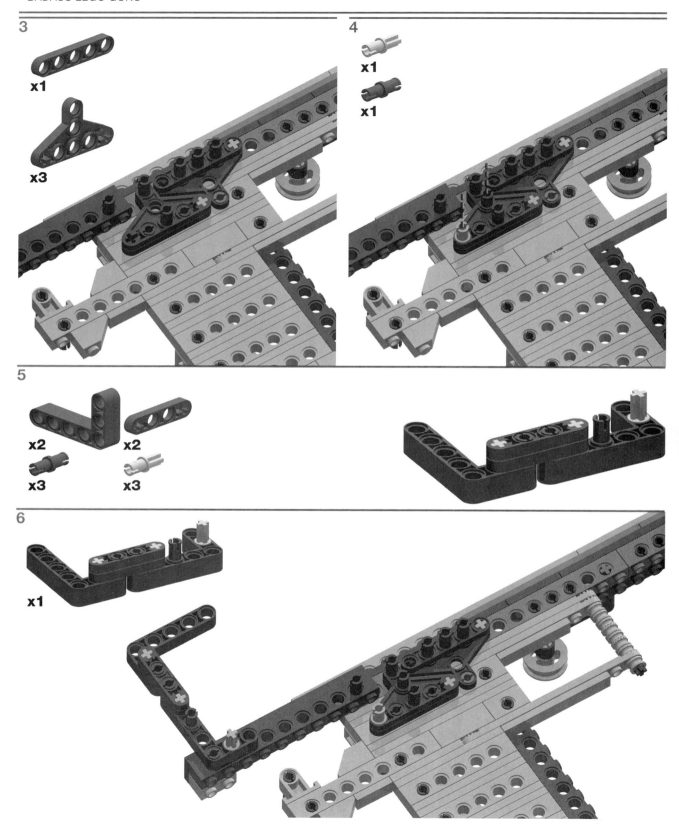

7

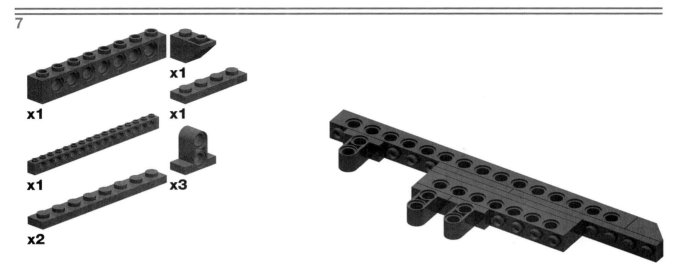

x1
x1
x1
x1
x3
x2

8

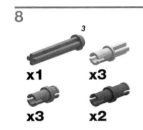

3

x1
x3
x3
x2

9

x1
x2

10

x1

11

8

x1

6

x1

x2

x2

12

x1

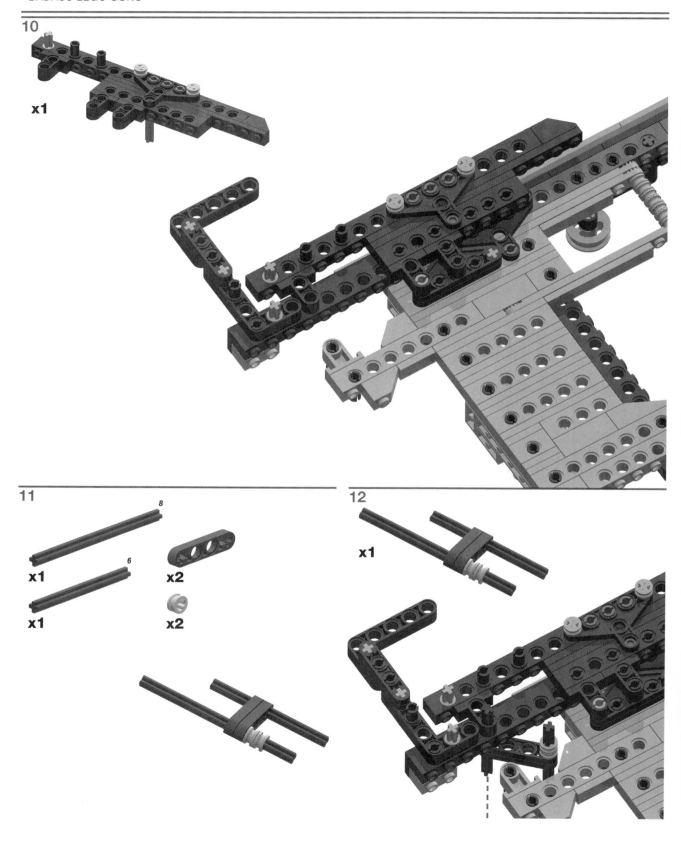

13

x1   x1

14

x1   x1

15

x1   x1   x1   x1

16

x1   x1   3   x4

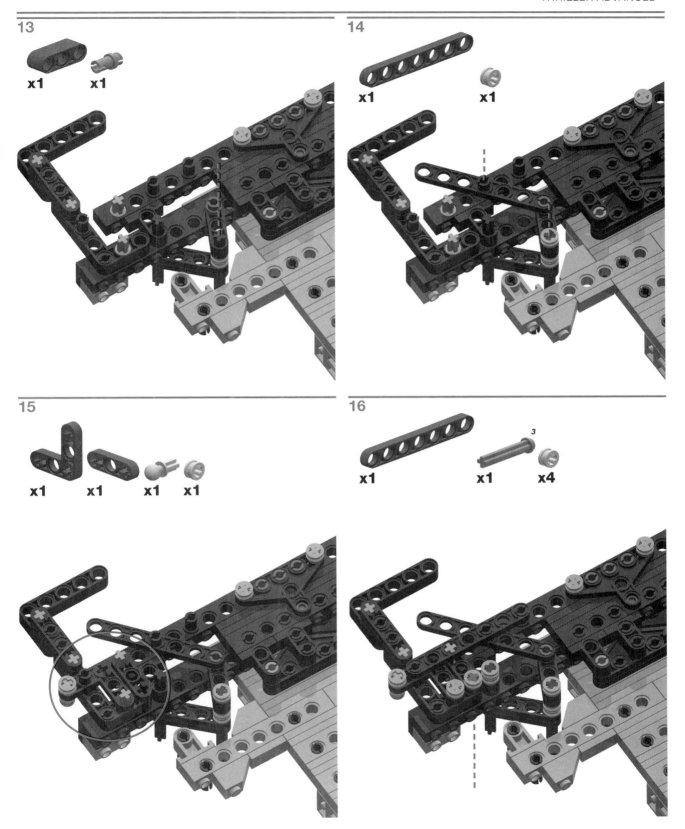

17

x1

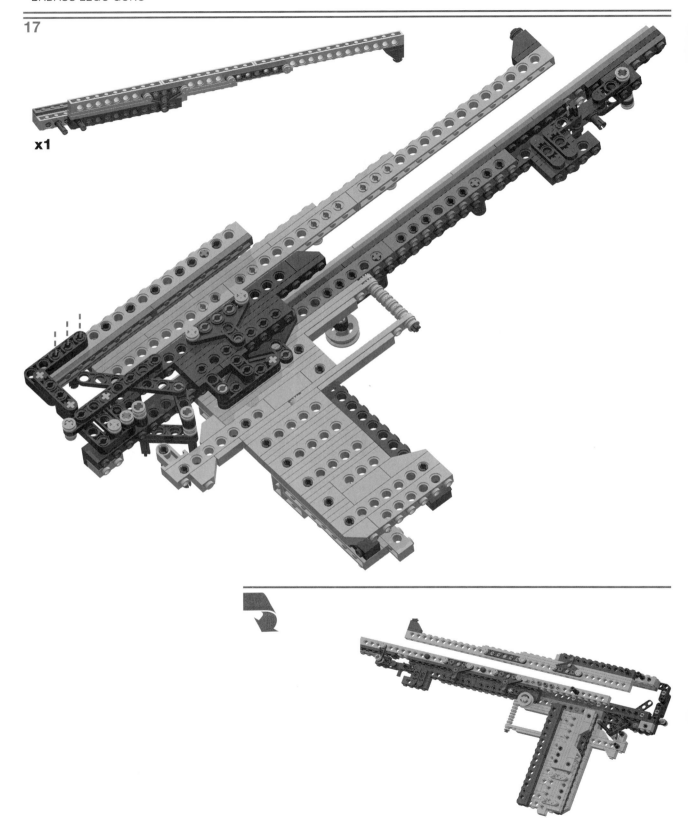

18

x2　x1

19

2

x2　x2

20

x2　x1

21

x2　x4

22

x1

23

x1

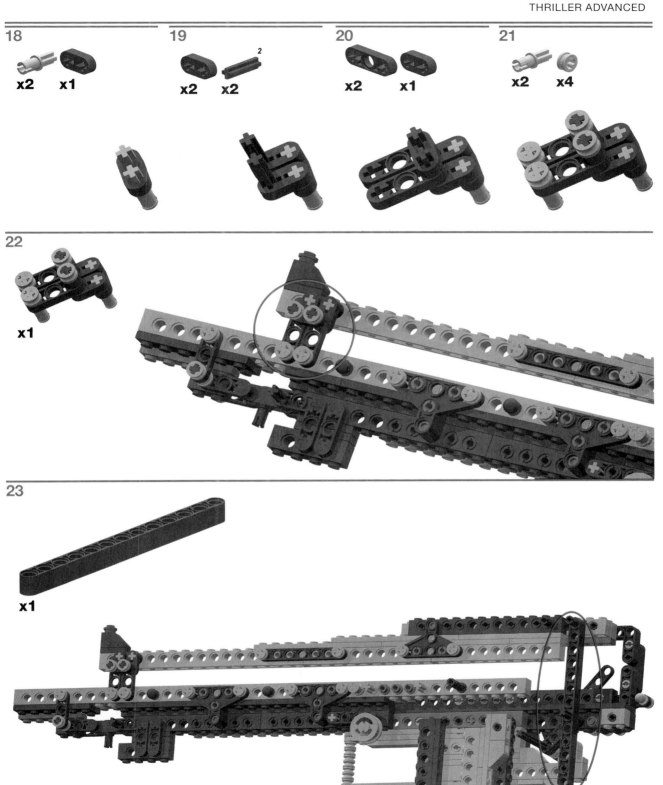

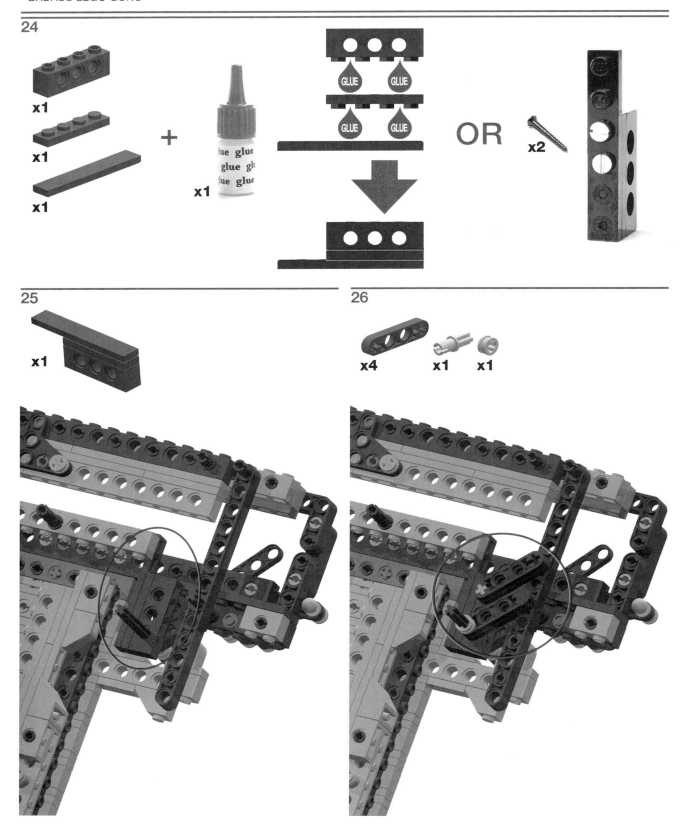

24

x1

x1

x1

+

x1

GLUE GLUE

GLUE GLUE

OR

x2

25

x1

26

x4 x1 x1

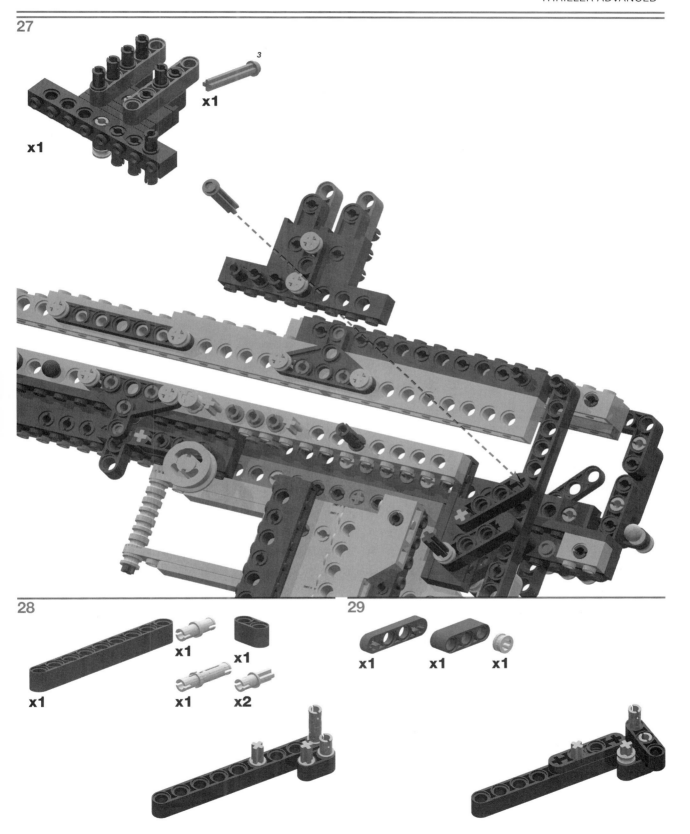

30

x1 **x1** **x1**

31

4

x1 **x1** **x1**

32

x1

33

x1 **x2**

x4

34

x1

35

x1 **x1**

3

x1 **x2**

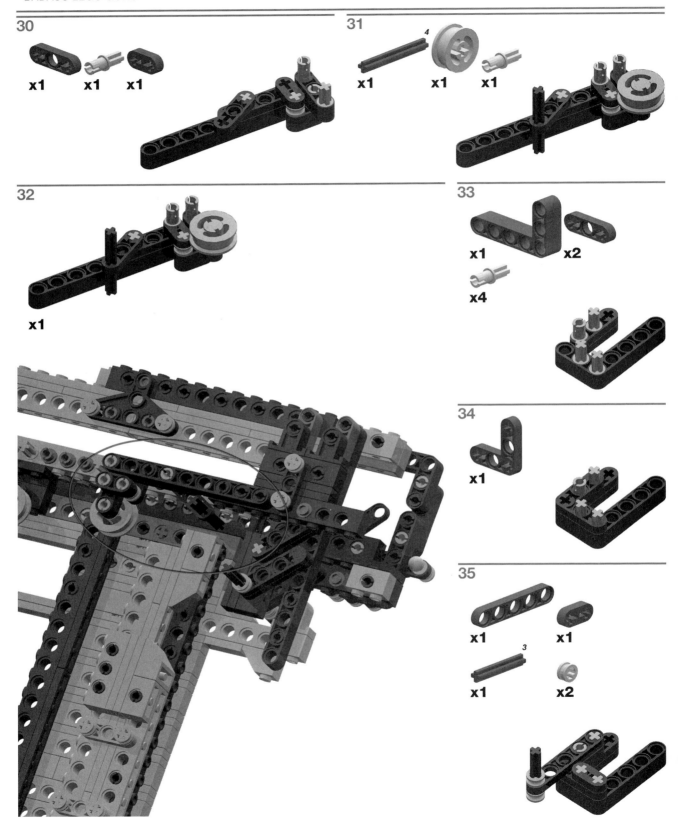

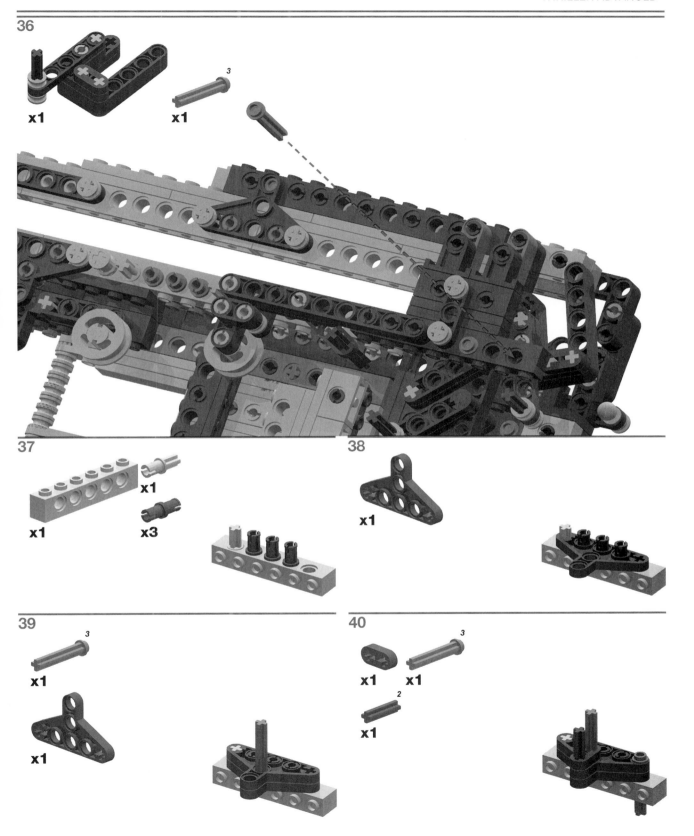

36
x1 x1 3

37
x1 x1
x3

38
x1

39
3
x1
x1

40
3
x1 x1
2
x1

41

x1

42

x1   **x2**

43

x1   **x1**

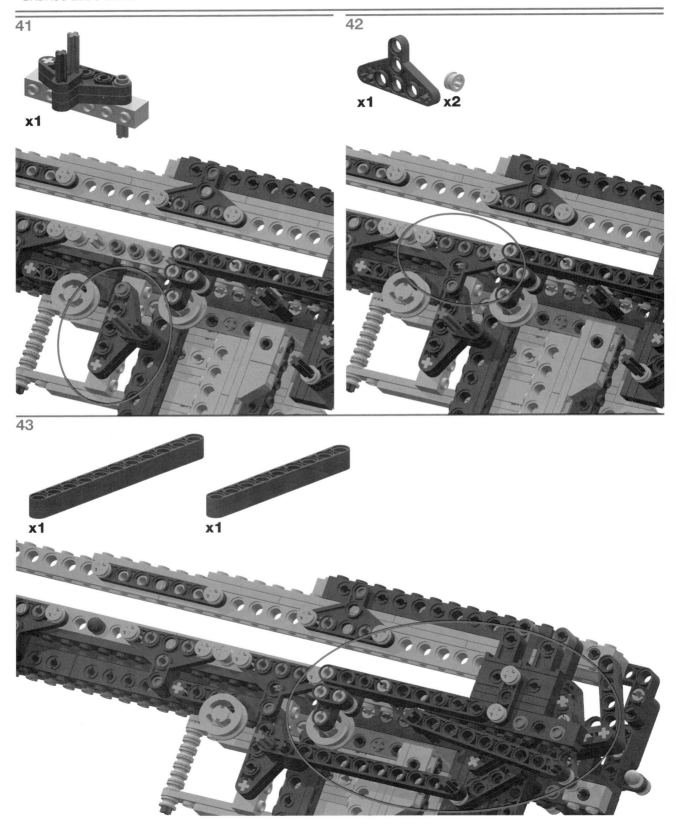

44

x1 **x5**

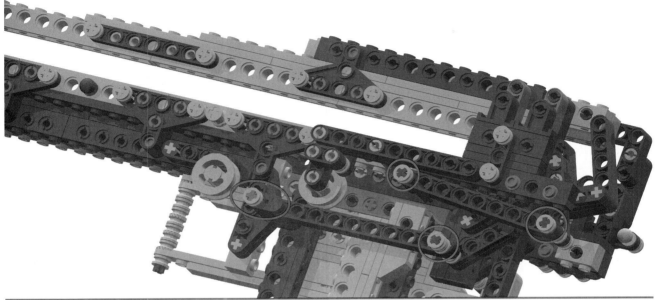

45

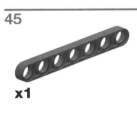

x1

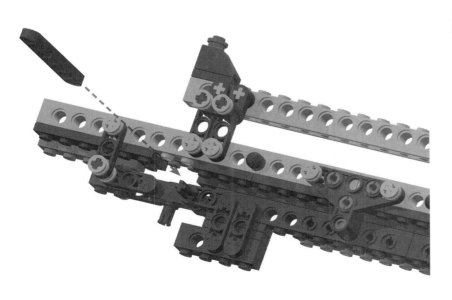

46

x2

47

x1

x1

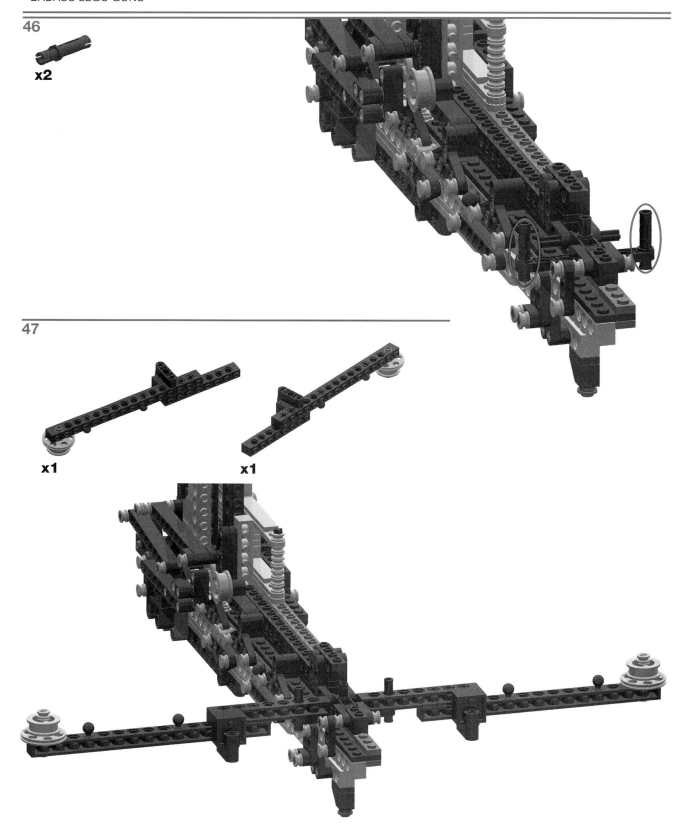

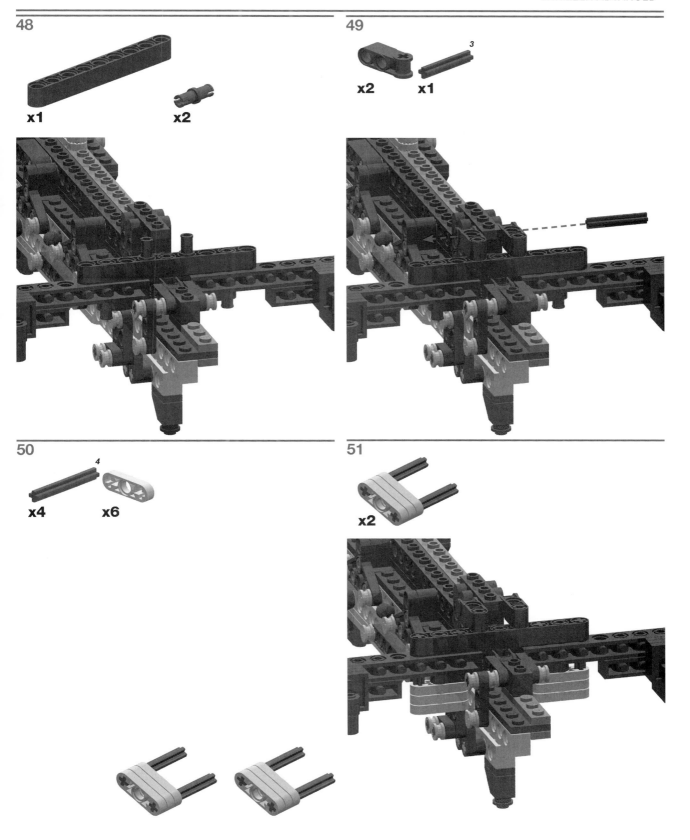

48

x1

x2

49

x2

x1

3

50

x4

x6

4

51

x2

52

x1

Make sure this face is sanded properly. Otherwise, the gun won't work.

Sandpaper

x1

53

x1

This modified Technic pin holds the projectiles inside the magazine.

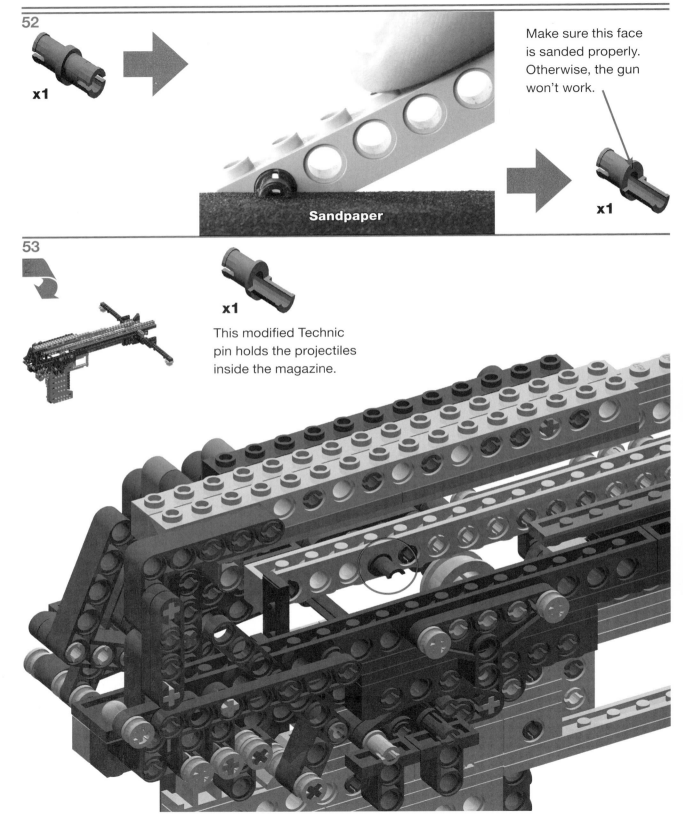

54

x1

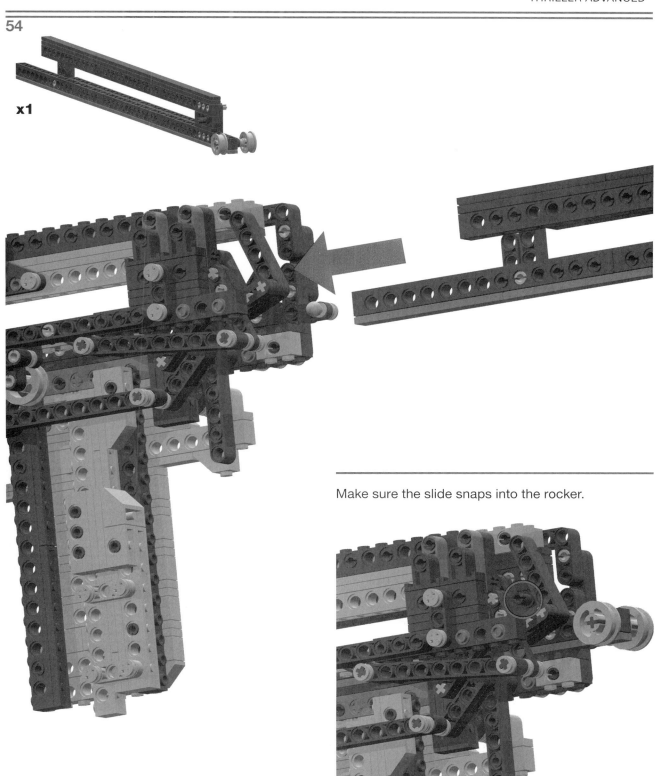

Make sure the slide snaps into the rocker.

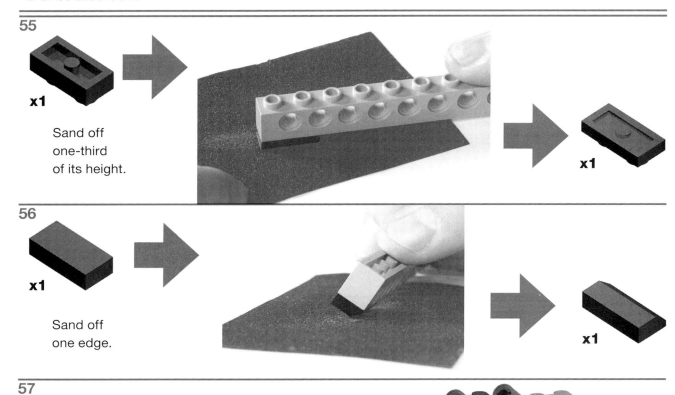

55

x1

Sand off
one-third
of its height.

x1

56

x1

Sand off
one edge.

x1

57

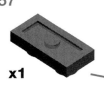

x1

This part holds the projectile
inside the launch rail.

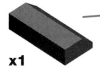

x1

This part is needed for the
ammunition to glide properly
into the launch rail.

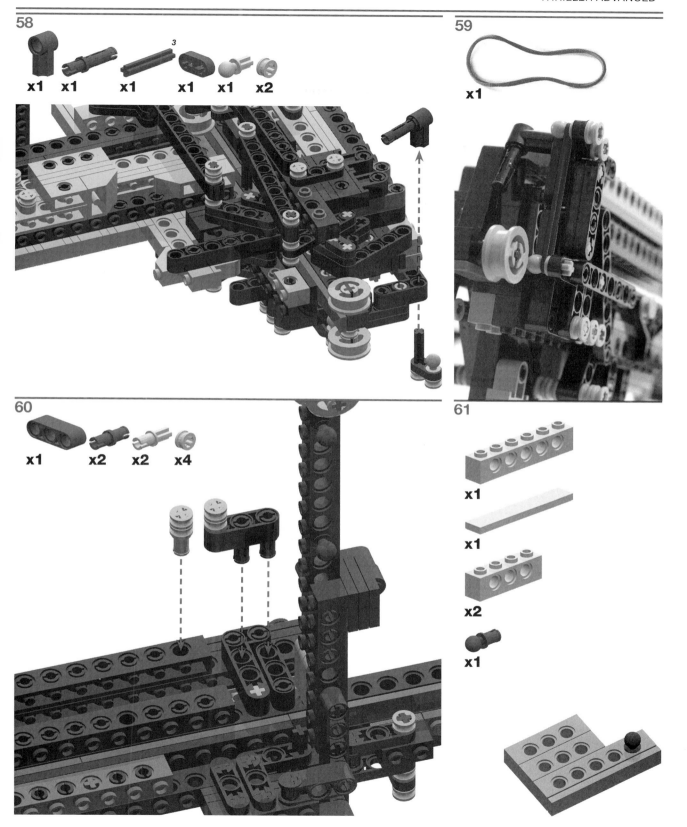

58

x1 x1 x1 x1 x1 x2

59

x1

60

x1 x2 x2 x4

61

x1

x1

x2

x1

ATTACHING THE RUBBER BAND

1

x1   **x1**

Installing the loading rubber band can be a bit tricky. Make sure it isn't too taut so that you can load nine projectiles.

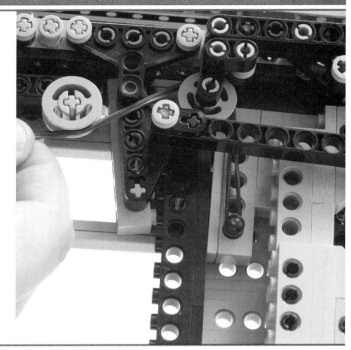

2

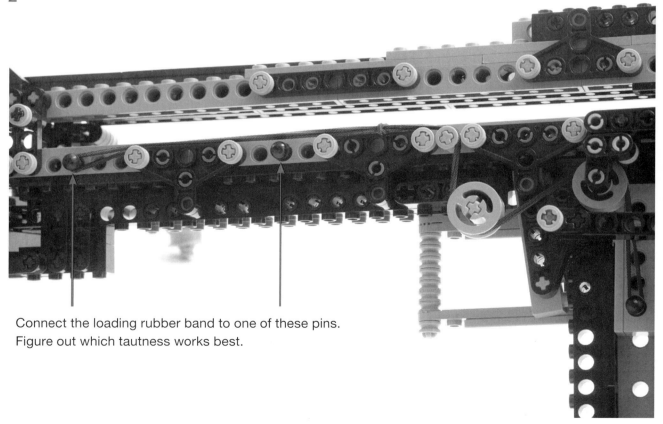

Connect the loading rubber band to one of these pins. Figure out which tautness works best.

LOADING THE GUN

1

x1
x1
x1

Tip: Gluing the parts together prevents your projectiles from bursting apart when shot at hard objects.

You will need at least nine projectiles.

2

Sometimes, your projectiles won't glide through the magazine properly, especially if your LEGO parts are new. Sanding them solves the problem.

Use fine sandpaper to prevent your LEGO parts from getting ugly scratches.

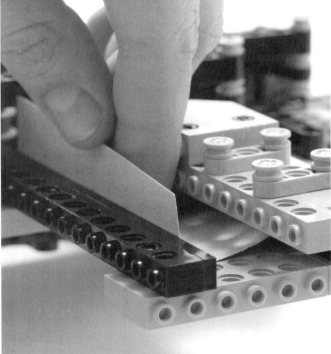

ATTACHING THE LAUNCHING BAND

1

x4

Connect two rubber bands in the manner shown here to make one launching band. Then, take two identical launching bands and use them as one.

2

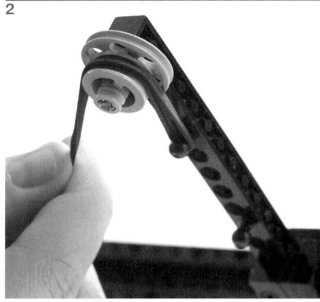

Take the launching band, fasten it to the pin, and pull it around the reel.

3

Stretch it, and then pull it through the slide and the launch rail.

4

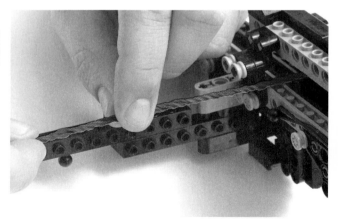

Now twist the launching band. This is very important for the trigger mechanism to function properly.

5

Pull it around the second reel, and attach it to the second pin.

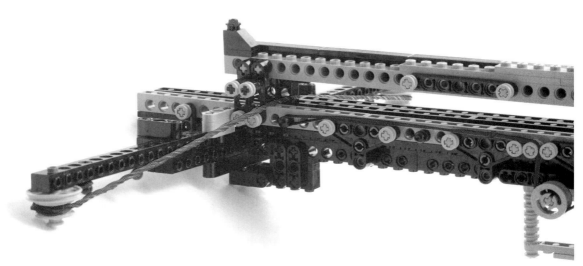

TIP: *You can alter the positions of the two pins to adjust the tension in the launching band.*

WARBEAST

★★★★★

FULLY AUTOMATIC ASSAULT RIFLE

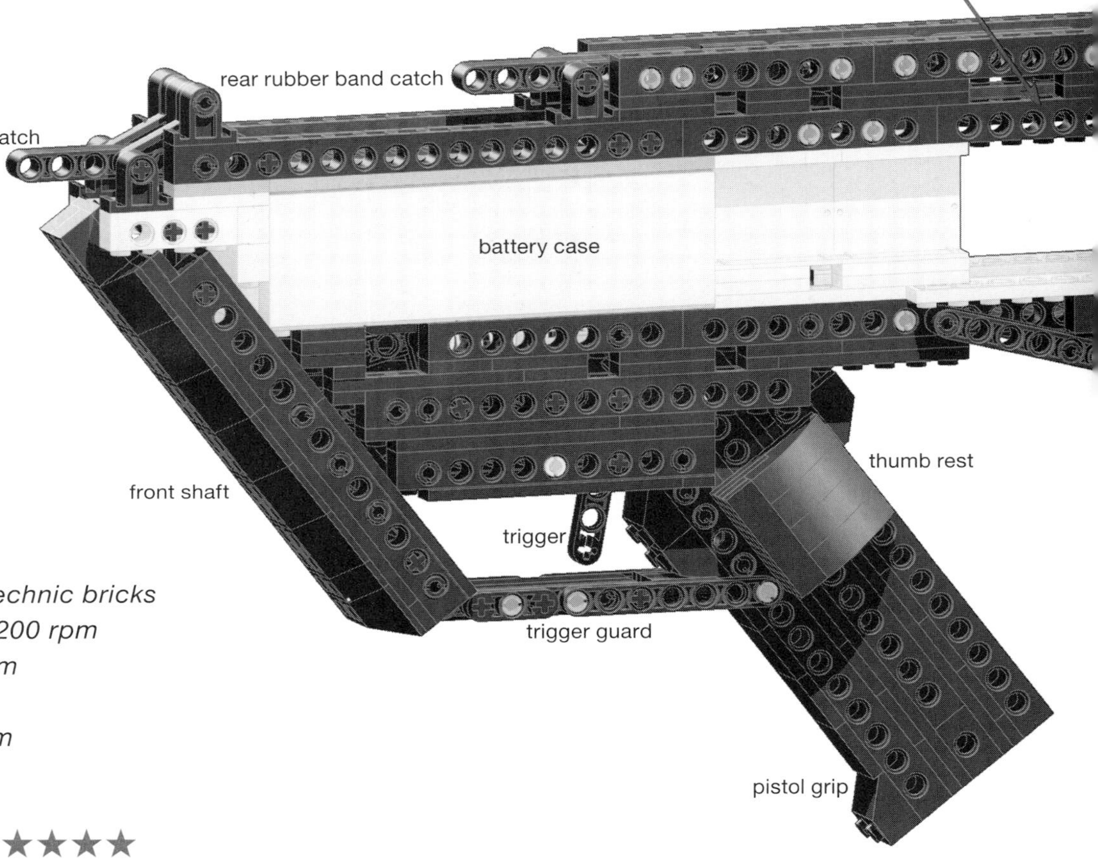

carrying handle

rear rubber band catch

front rubber band catch

battery case

front shaft

thumb rest

trigger

trigger guard

pistol grip

AMMO: *30 2×1 Technic bricks*
RATE OF FIRE: *1200 rpm*
LENGTH: *51.71 cm*
WIDTH: *5.56 cm*
HEIGHT: *30.51 cm*
WEIGHT: *1200g*
PARTS: *895*
FIRE POWER: ★★★★★
LEVEL: *Expert*

The Warbeast projectile

Carrying the Warbeast by its handle

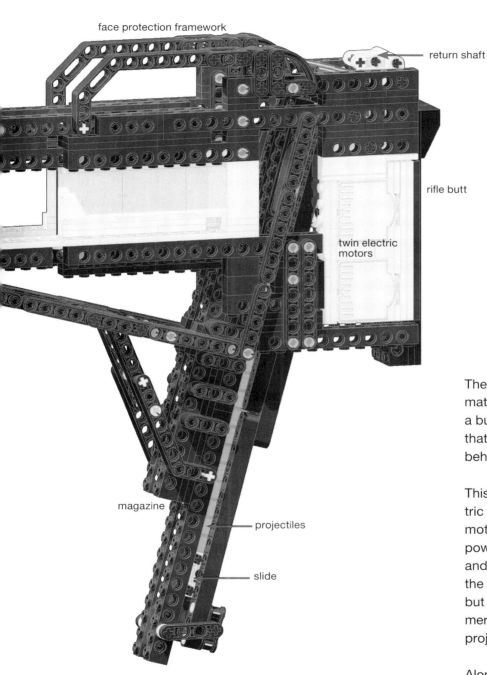

face protection framework

return shaft

rifle butt

twin electric
motors

magazine

projectiles

slide

The Warbeast is a fully automatic assault rifle based on
a bullpup design. This means
that the magazine is located
behind the pistol grip.

This Warbeast is the only electric model in this book. LEGO
motors don't have enough
power to stretch rubber bands
and are generally unsuitable for
the construction of weapons,
but in the Warbeast, they are
merely responsible for firing the
projectiles at a consistent rate.

Along with the Thriller Advanced,
the Warbeast spent a lot of time
in development. Its large size
allows for many design variations and options.

WARBEAST DESIGN HISTORY

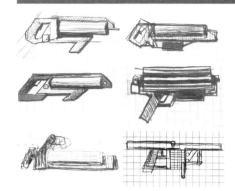

The goal behind the Warbeast was to construct a brick-shooting, fully automatic rifle with a 30+ round magazine and a rate of fire of more than 500 rounds per minute.

My original idea was a Gatling design, but that proved impractical. It would have made the gun too large and bulky. My second idea was to create a rifle with a large magazine driven by one large rubber band. That didn't work either.

The breakthrough idea was to use Technic 2×1 bricks connected to rubber bands as projectiles. These assemblies would be slipped into the magazine under tension and forced out by motor power.

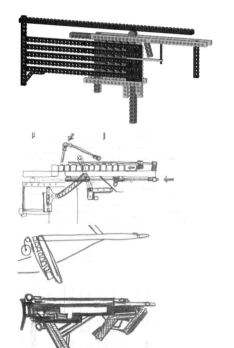

It was tough to find the right angle of incline for the magazine. If the projectiles were too tight in the magazine, they might get stuck; too loose, and they could slip out on their own. Equally challenging was finding a stable shape without the Warbeast becoming too clunky. Large models have far more possible variations than smaller ones, which is why the Warbeast took the longest to develop of all my guns.

To make the Warbeast as compact as possible, I decided on a bullpup design. In a bullpup design, the magazine is located behind the pistol grip, rather than in front of it (as in most rifles). This design allows for a long barrel length in a shorter firearm. The gun's mechanisms are located in the rifle butt, which is unused space in conventional rifles.

The Warbeast's theoretical rate of fire is 1,200 rounds per minute! If you don't need quite that speed, you can use just one motor instead of two.

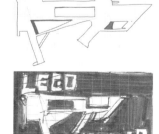

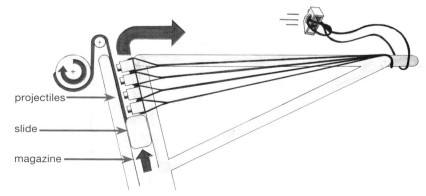

projectiles

slide

magazine

1

2

3

4

5

6

7

8

9

10

11

12

13 FINAL VERSION

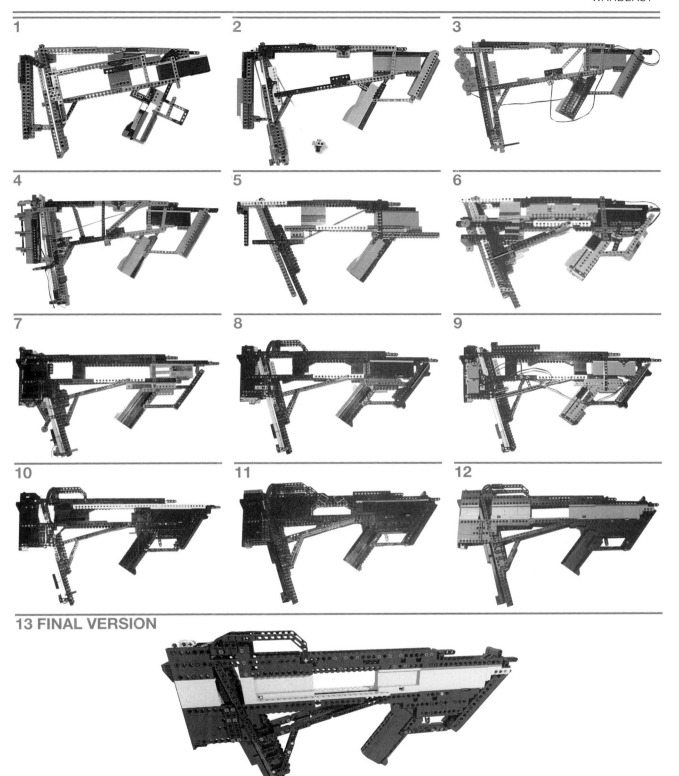

HOW IT WORKS

THE TRIGGER MECHANISM

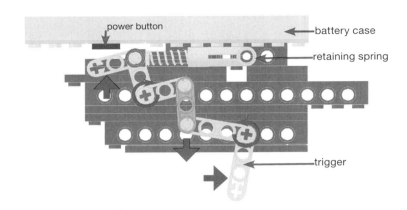

power button

battery case

retaining spring

trigger

THE FIRING MECHANISM

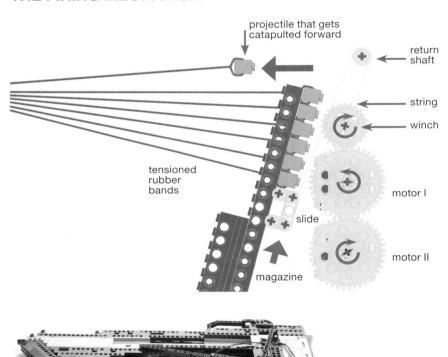

projectile that gets catapulted forward

return shaft

string

winch

tensioned rubber bands

motor I

slide

motor II

magazine

The Warbeast, loaded with 20 rounds

The Warbeast projectile is a 2×1 Technic brick attached to a rubber band. To load the weapon, hook the projectile's rubber band into one of the catches. (Depending on the strength of the rubber band, you can use the front or the rear catch.) Next, slip the projectile into the magazine, pulling the rubber band taut.

Pulling the trigger pushes the button on the 9V battery pack and powers the two electro-motors that drive a winch. This winch pulls up the slide inside the magazine, which pushes the projectiles out of the top of the magazine. Because their rubber bands are under tension, the projectiles instantly catapult out of the weapon at high speed.

Assuming the batteries are full and two motors are used, the slide will move so quickly that the entire 30-round magazine will empty in 1.5 seconds. This is equivalent to a firing rate of 1,200 rounds per minute! I recommend short bursts of fire.

BILL OF MATERIALS

x2 x2 x16 x2 x4 x2 x4 x4 x12 x1 x8 x4 x16 x10 x2 x4 x1 x2 x1 x2 x3 x19 x2 x2 x1 x3 x4 x3

x2 x6 x2 x2 x2 x10 x2 x3 x2 x4 x2 x4 x6 x4 x12 x2 x6 x8 x8 x4 x7 x21 x4 x3 x3 x7 x1 x1

x3 x1 x1 x4 x2 x1 x2 x4 x4 x2 x2 x35 x2 x5 x4 x30 x13 x2 x17 x4 x14 x6 x8 x24 x5 x13 x5 x9 x2 x2 x1 x2 x2 x2

x3 x1 x11

x2 x21 3 x7 x4 x15 5 x2 6 x3

x8 x64 x22 x136 x48 x1

x2 x1

Micromotor
x10 x3 x1 pulley

x1

x2 x1

Length:
69 studs

One piece of
sandpaper

Length:
15 studs x1

FOR THE PROJECTILES

Several rubber bands,
preferably 85 mm in
diameter

x30

20 inches of thread
(or dental floss)

x6

NOTE: *1.2V NiMh AA batteries are
rechargeable, but 1.5V alkalines
make the Warbeast fire more rapidly.*

PISTOL GRIP

1

x1

x1

x4

2

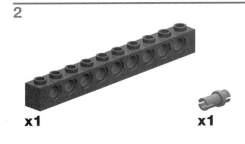

x1

x1

3

x2

4

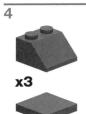

x3

x4

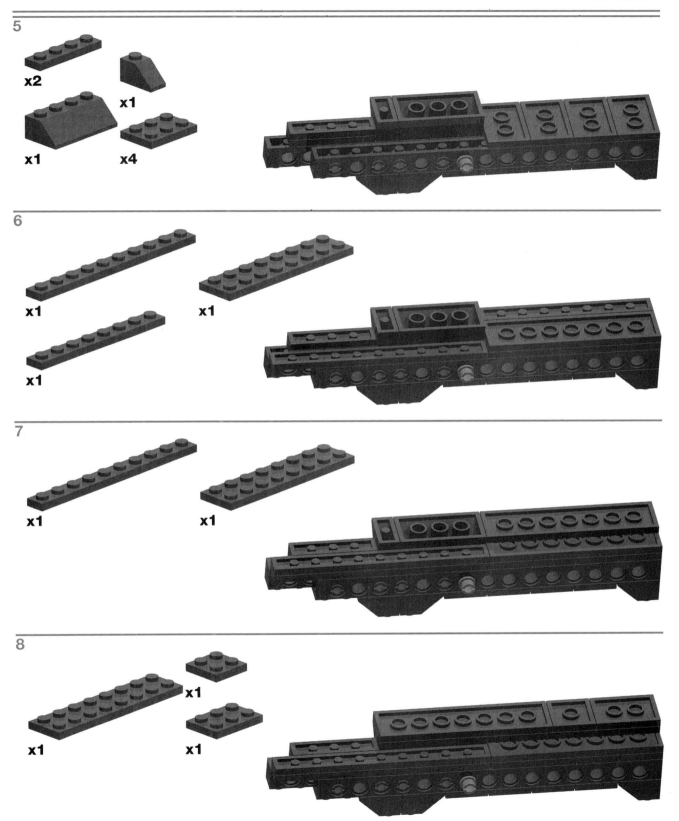

9

x1

x1

x1

10

x1

x1   **x1**

11

x2

x2

x1

x1

12

x1   **x6**

x1

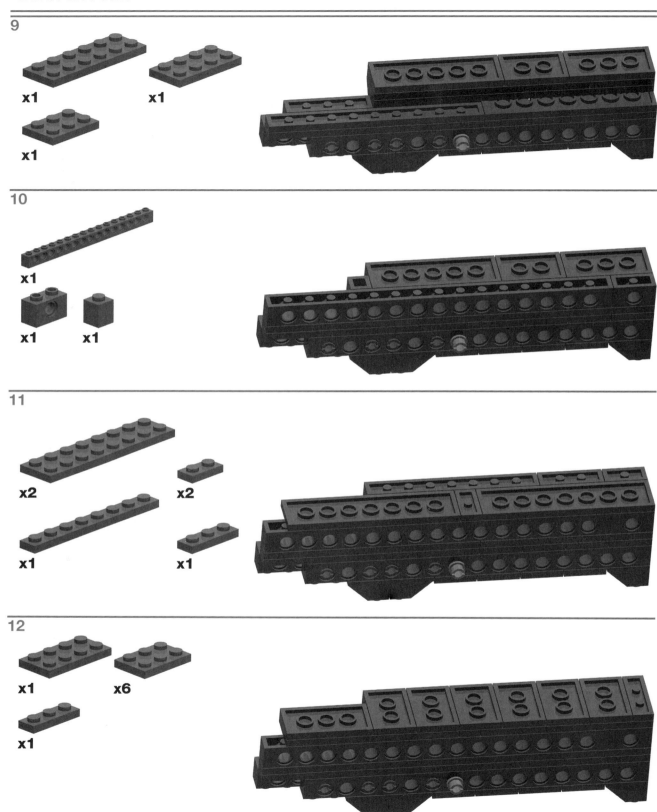

13

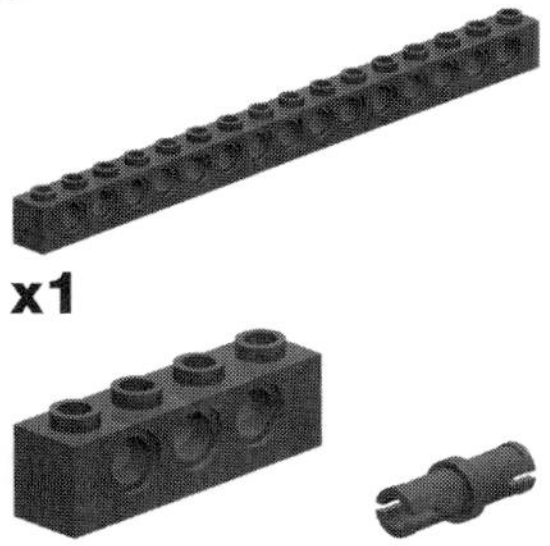

x1

x1   x1

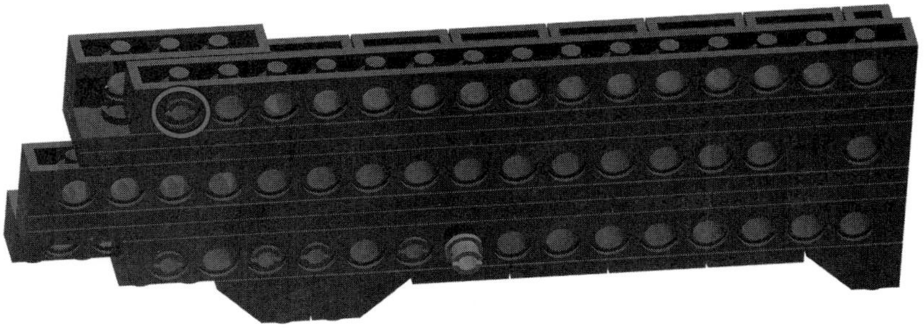

14

x6   x1   x1

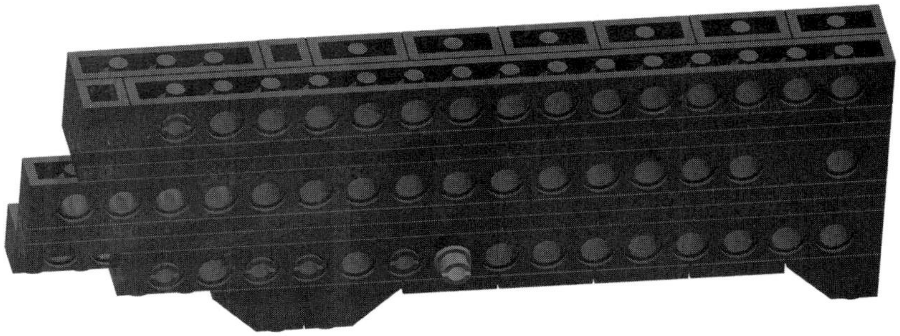

15

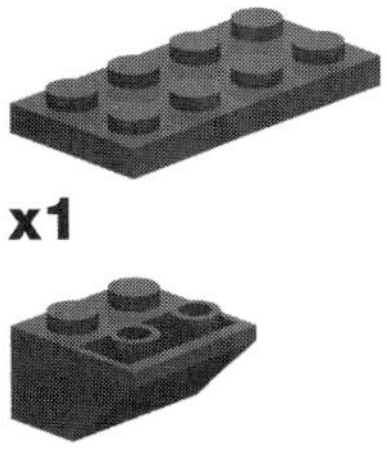

x1

x1

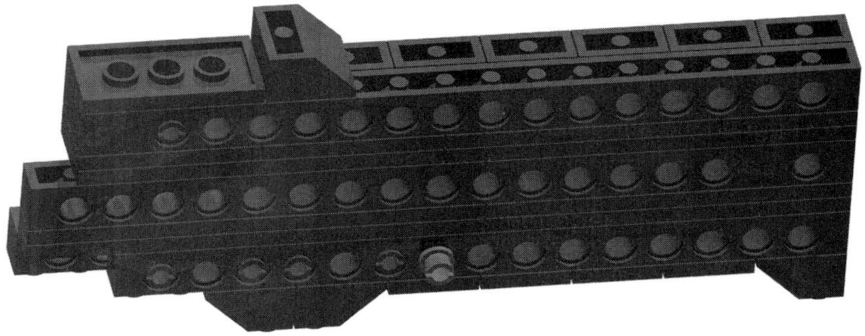

16

x2

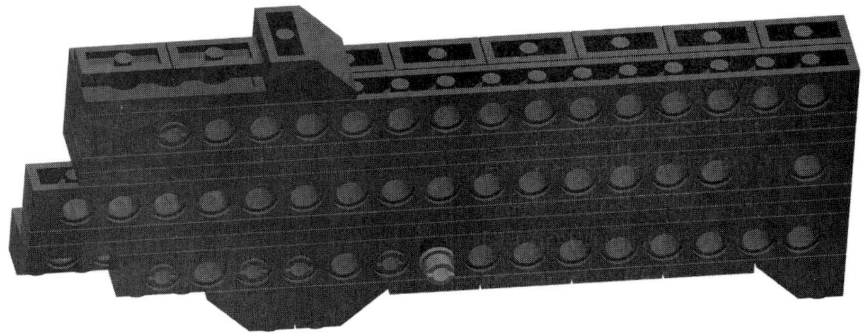

17

x1

18

x8

19

x2

20

x1 **x1** **x1**

21

x4

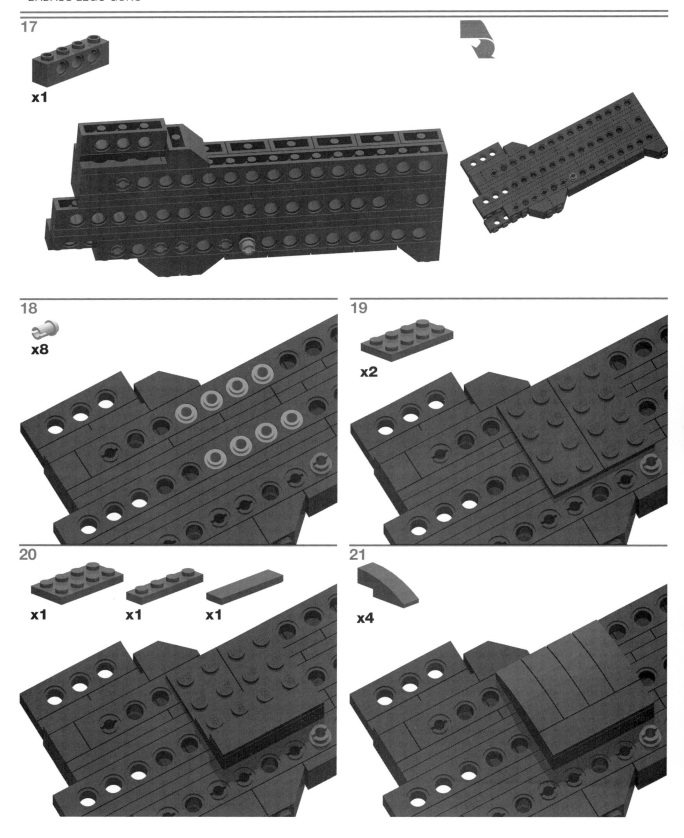

RIFLE BUTT

1

x1 **x4**

2

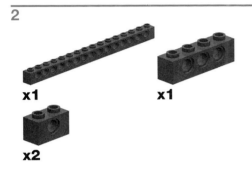

x1 **x1**

x2

3

x1 **x3**

4

x2

5

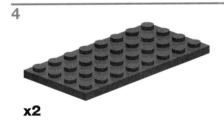

x1

x1 **x2**

6

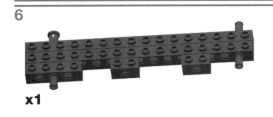

x1

7

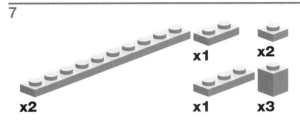

x2   x1   x2   x1   x3

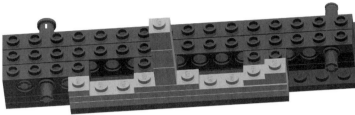

8

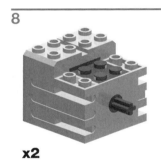

x2

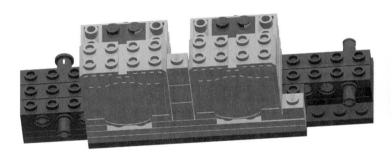

NOTE: *If you don't have a second motor available, no problem: the Warbeast only needs one. The second one only increases the rate of fire. Just leave out the left motor, and fill the gap with gray bricks.*

9

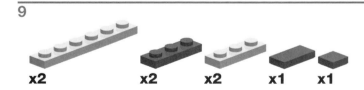

x2   x2   x2   x1   x1

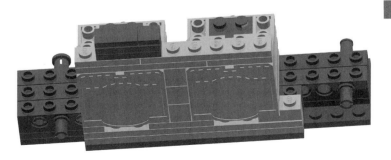

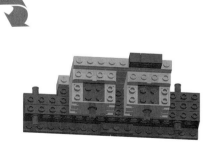

10

x2

11

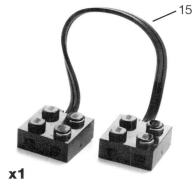

15 studs length

x1

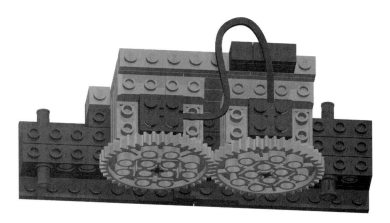

12

x2

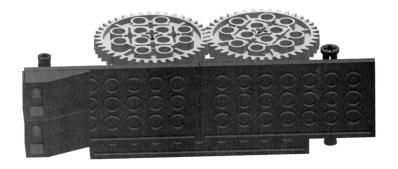

TRIGGER

1

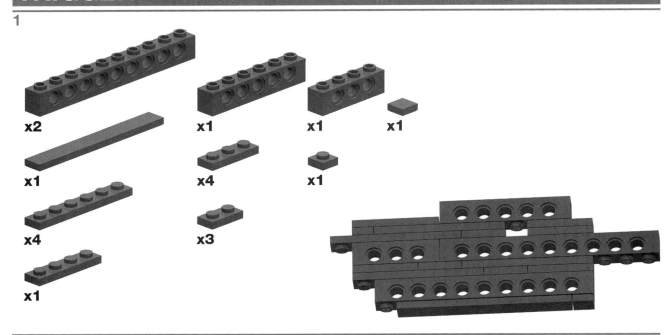

x2

x1

x1

x1

x1

x4

x1

x4

x3

x1

2

x2

x6

x1

3

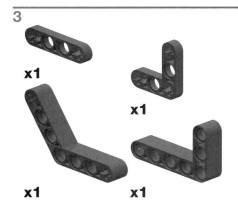

x1

x1

x1

x1

4

x1 x6 x1

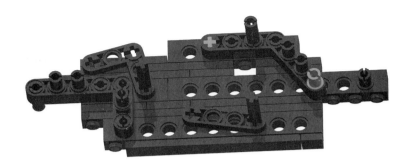

5

x1

x1

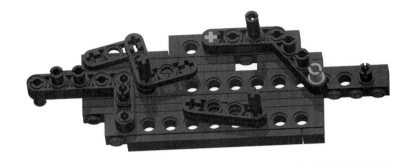

6

x3 x1

7

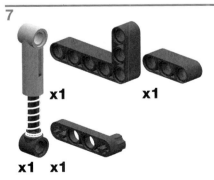

x1 x1

x1 x1

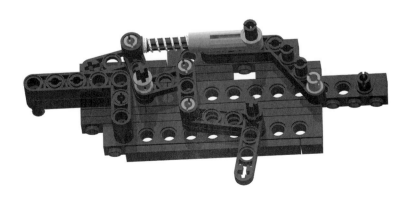

8

x4

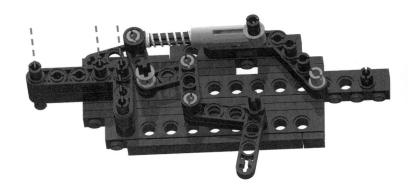

BARREL

1

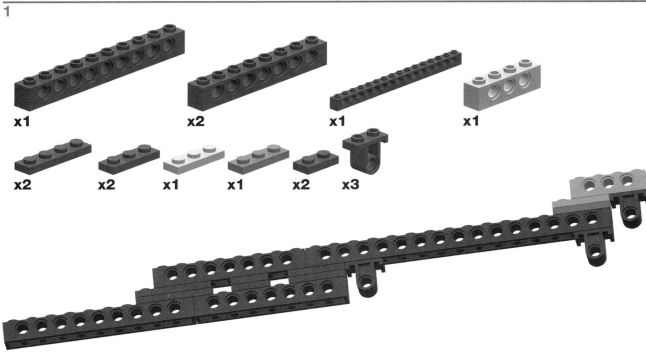

x1 x2 x1 x1

x2 x2 x1 x1 x2 x3

2

x9

3

x2 **x1** **x1**

4

x1 **x2** **x1** **x1**

x2 **x2** **x1** **x1** **x2** **x3**

5

x9

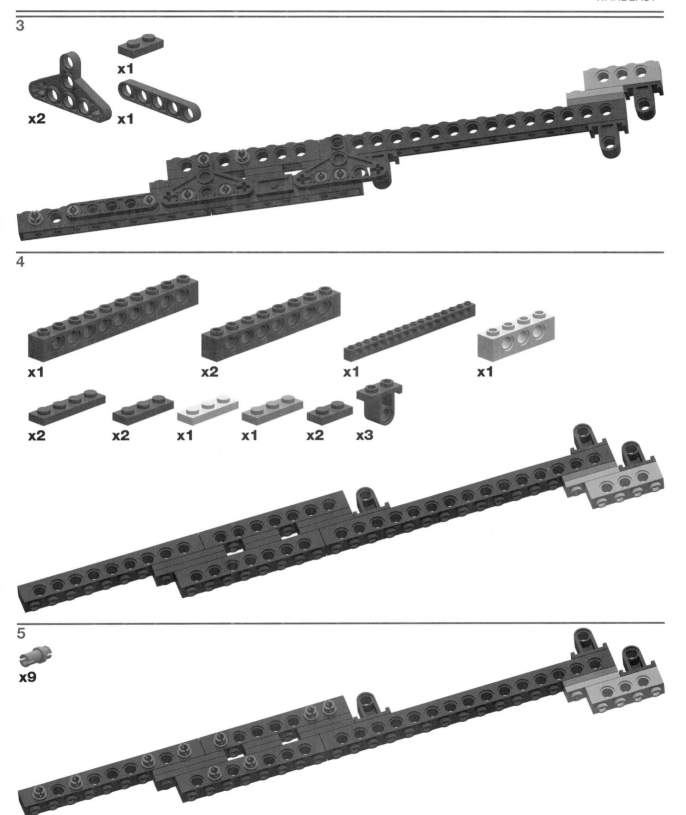

6

x7

4

7

x1 x2

x3 x1

x1

8

x1 x2 x2

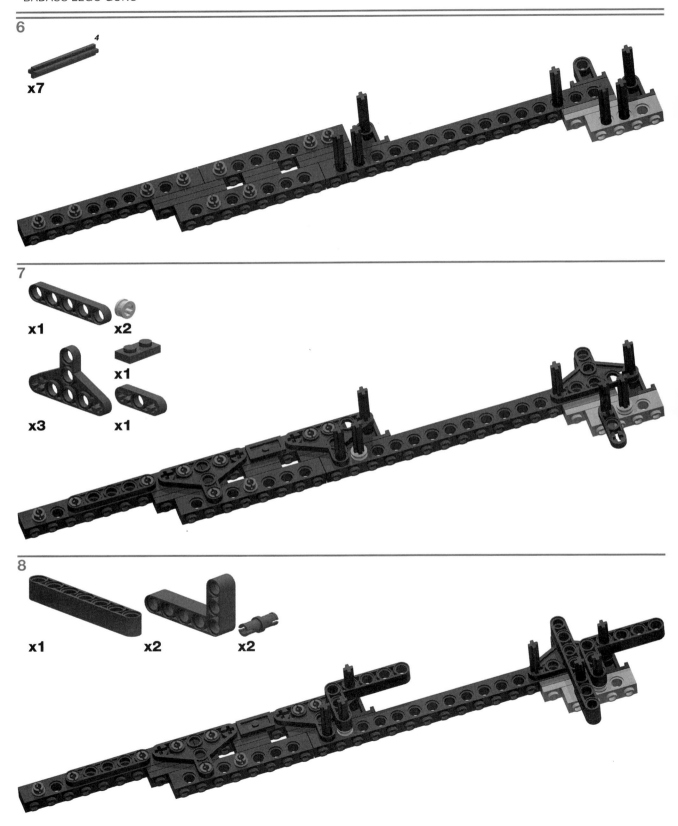

9

x5

10

x1

11

x1   **x2**   **x1**

12

x1

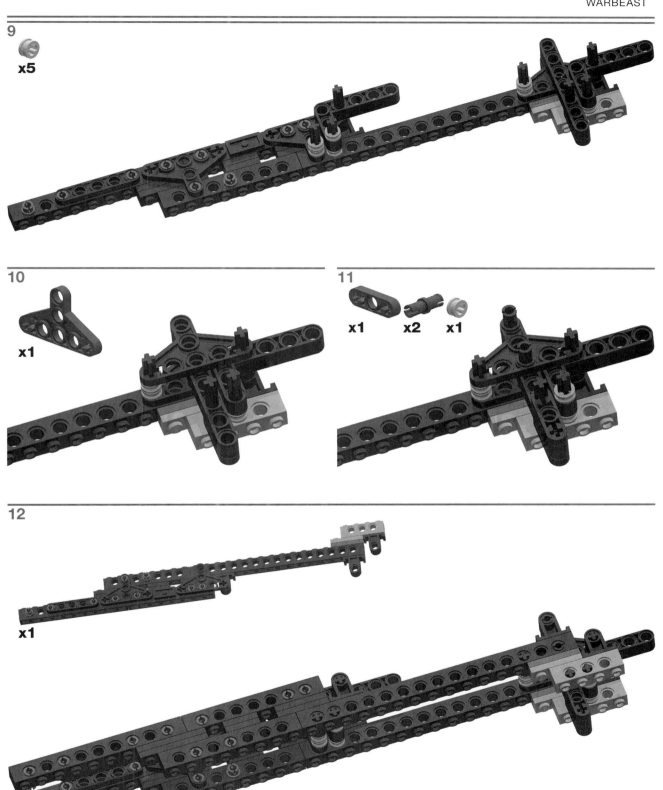

13

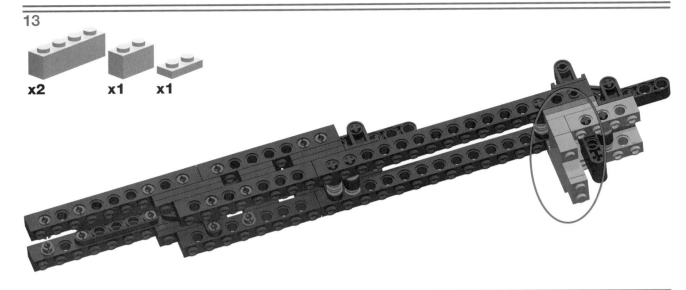

x2   x1   x1

MAGAZINE: LEFT SECTION

1

x1   x2   x3   x1   x2   x1

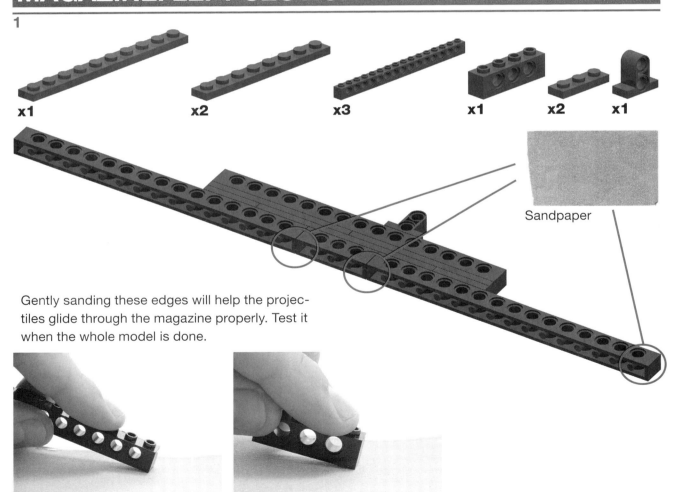

Sandpaper

Gently sanding these edges will help the projectiles glide through the magazine properly. Test it when the whole model is done.

2

x15 **x8** **x1**

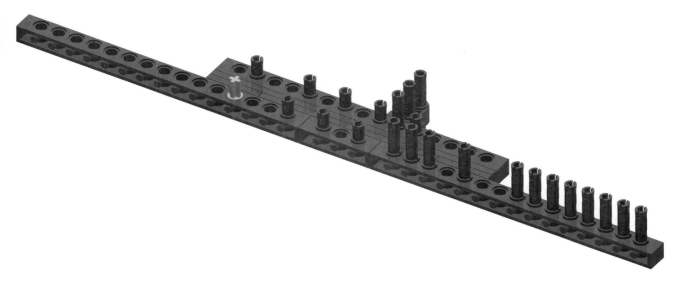

3

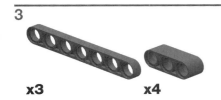

x3 **x4**

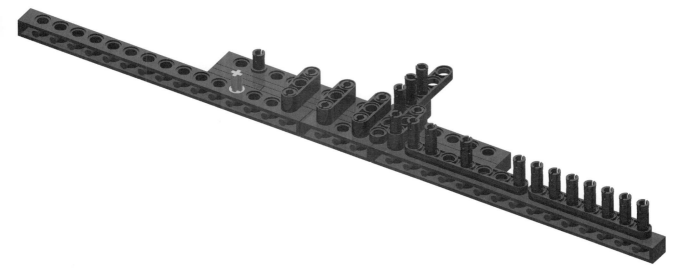

4

x1

x1

3
x1

x2

x2

x1

5

x1

x2

6

x1

x1

x1

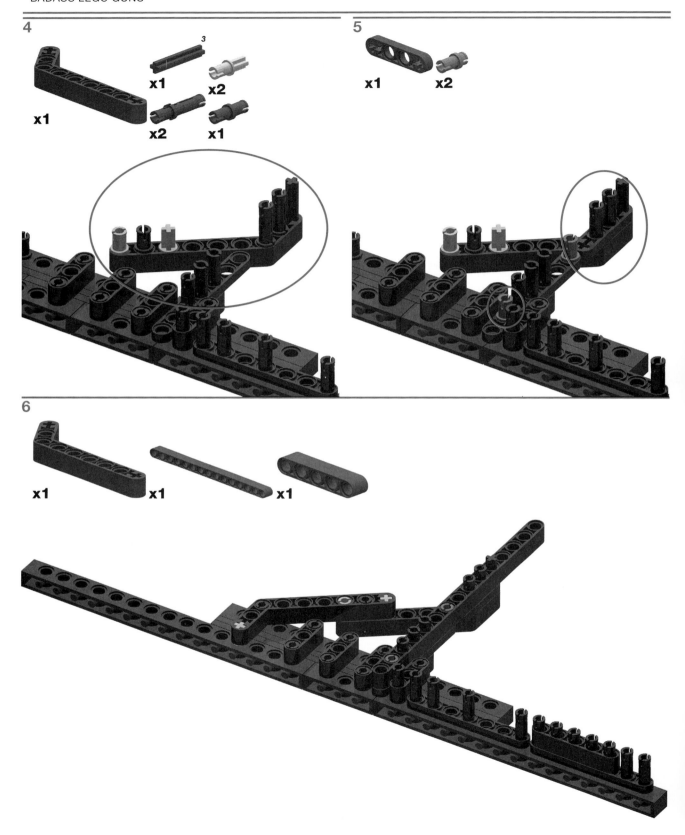

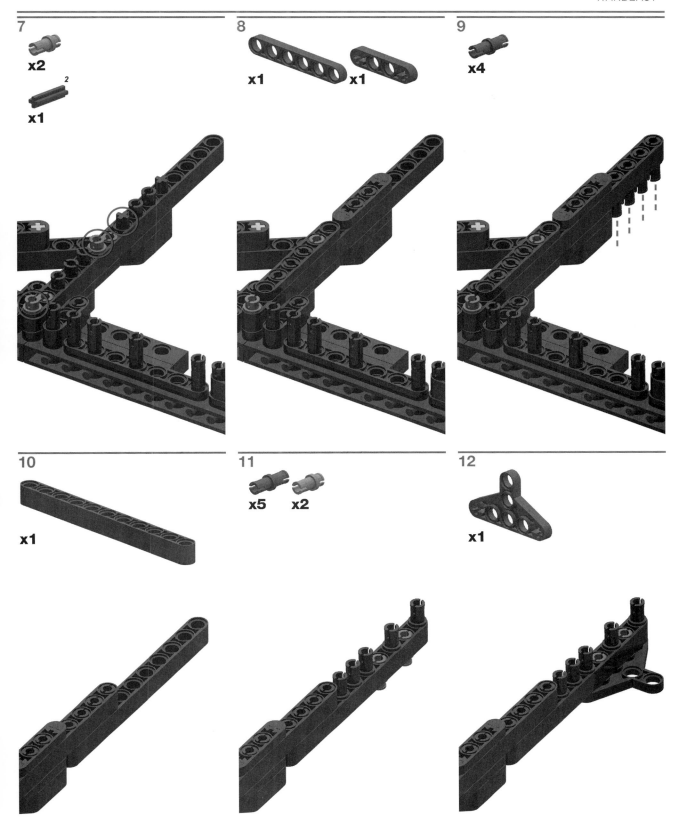

MAGAZINE: CENTER SECTION

1

x2

x2

x2

x4

x18

x8

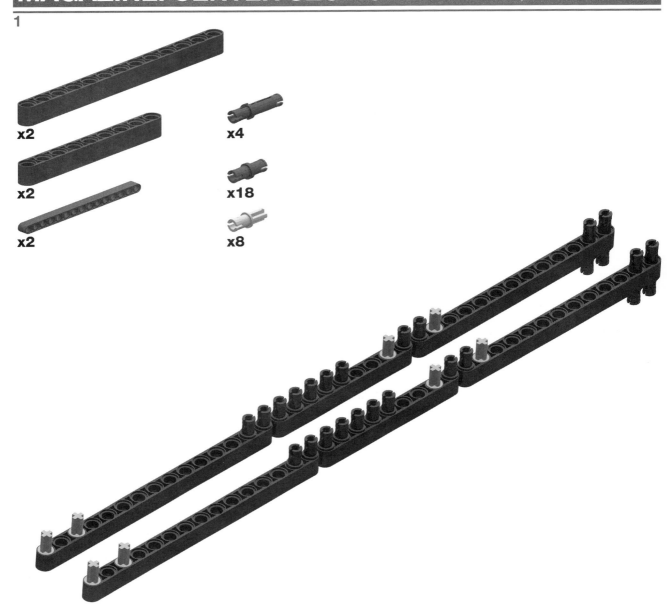

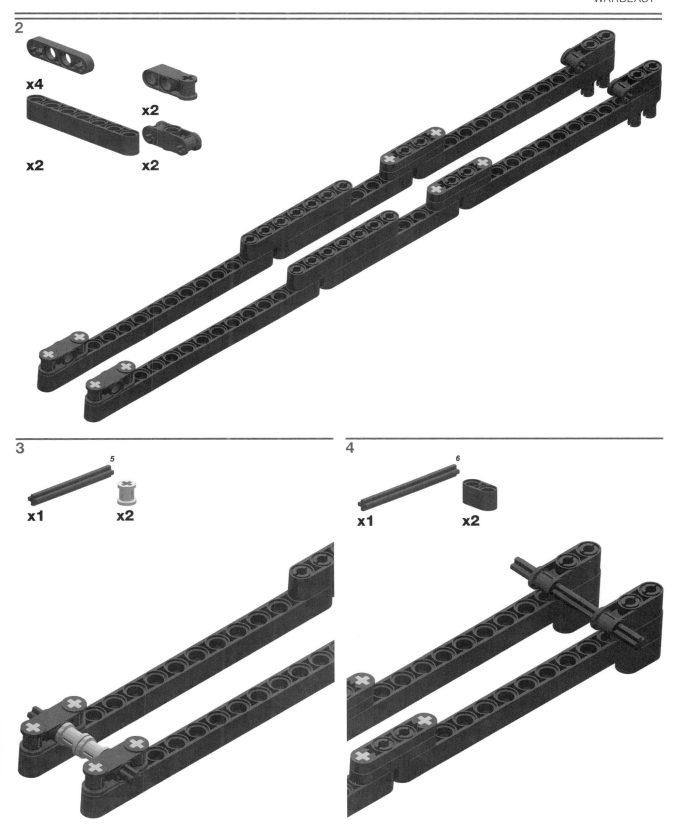

MAGAZINE: RIGHT SECTION

1

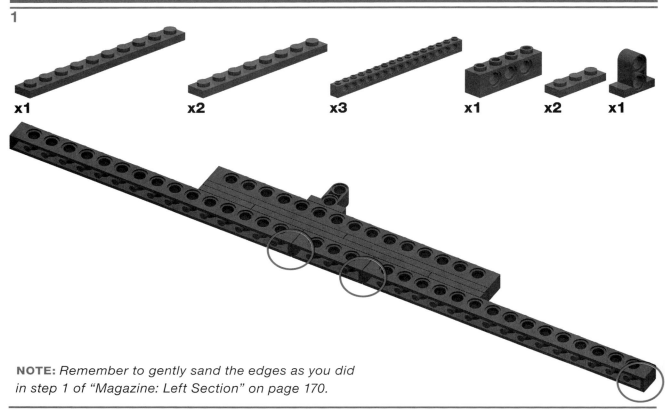

x1   x2   x3   x1   x2   x1

NOTE: *Remember to gently sand the edges as you did in step 1 of "Magazine: Left Section" on page 170.*

2

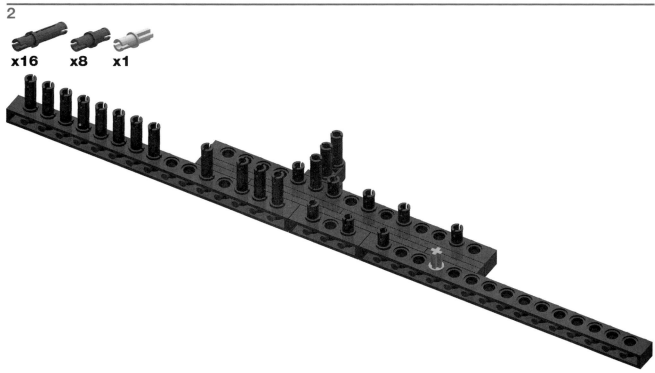

x16   x8   x1

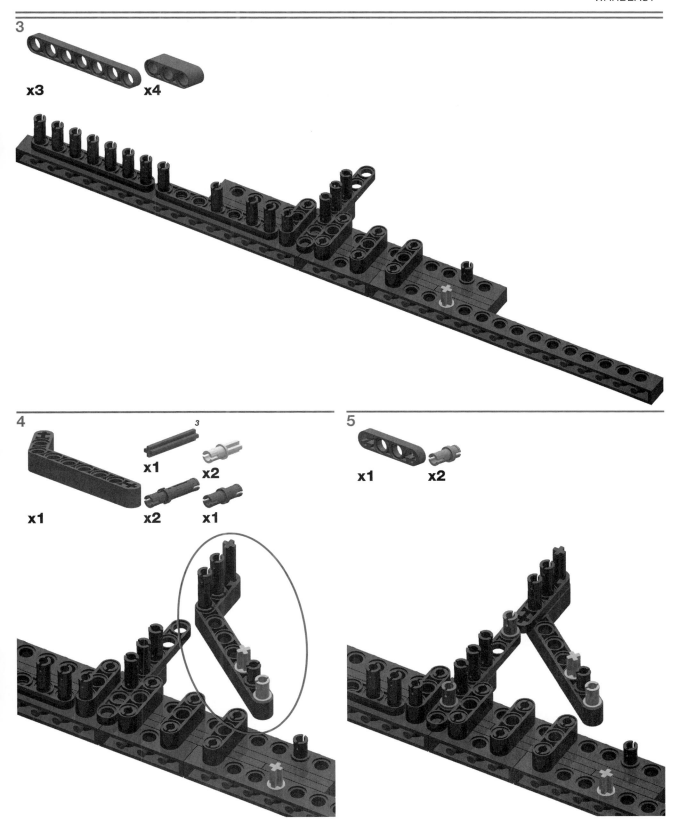

6

x1

x1

7

x2 2

x1

8

x1 x1

9

x4

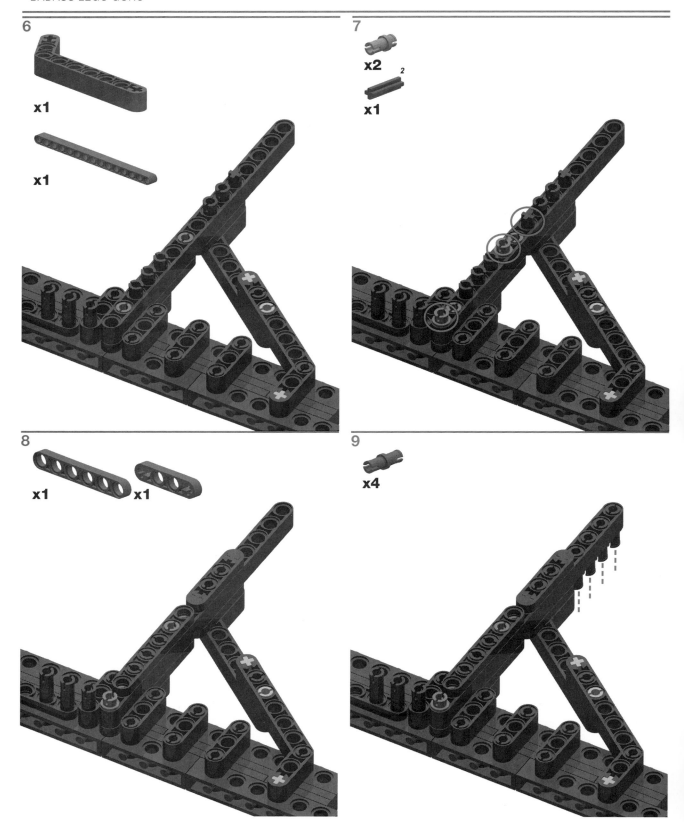

10

x1

11

x5   **x2**

12

x1

13

x1

14

x2

x1

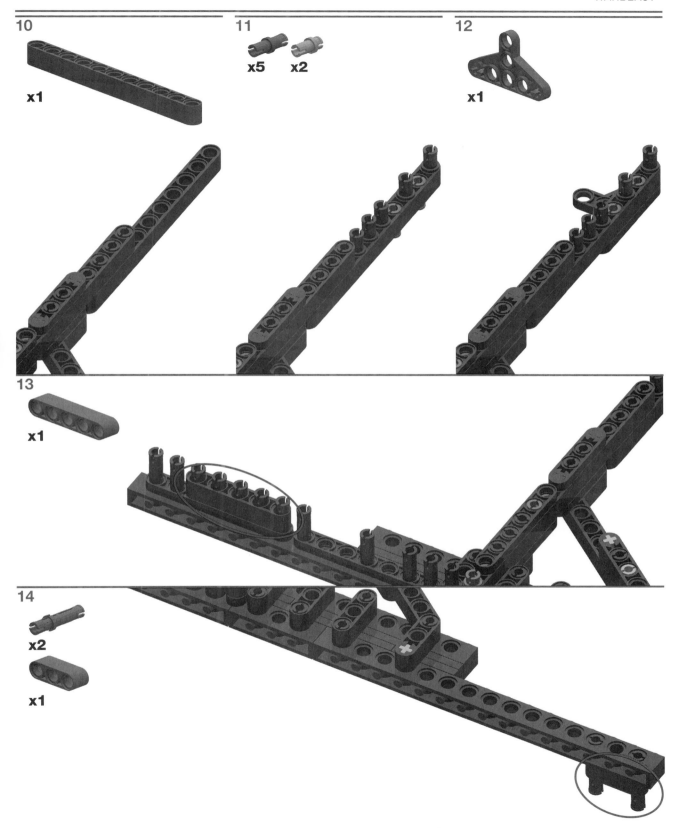

BODY: LEFT SECTION

1

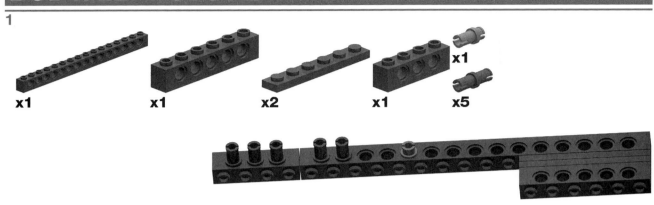

x1 x1 x2 x1 **x1**

x5

2

x1

x1

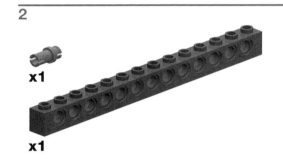

3

x1 x1 **x2**

x4

4

x1

x1

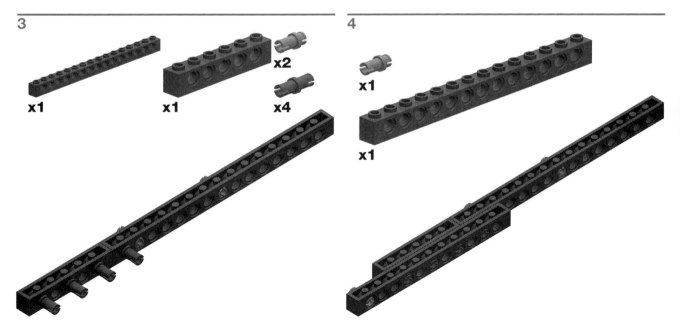

5

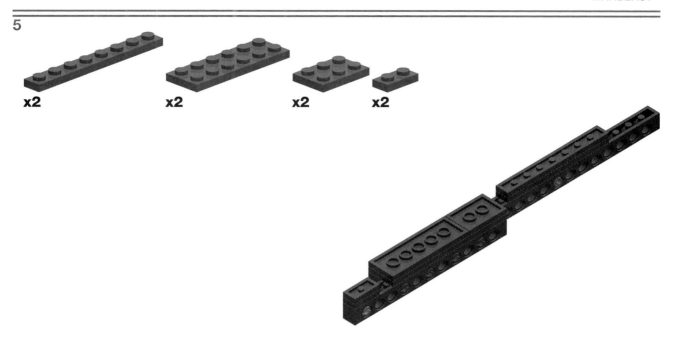

x2   x2   x2   x2

6

x1

7

x14

x1

x1

x1

8

x1

x1

x1

x1

x1

9

x1

x1

x1

x1

x1

x1

x1

x1

x4

x2

x1

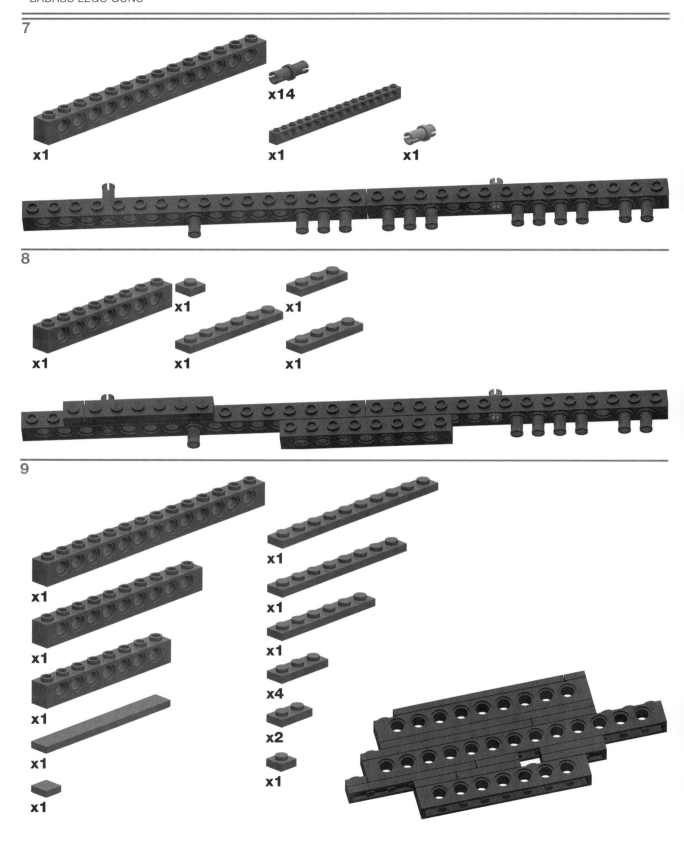

10

x2 x1 x1

11

x2

12

x1 x13 x1 x2

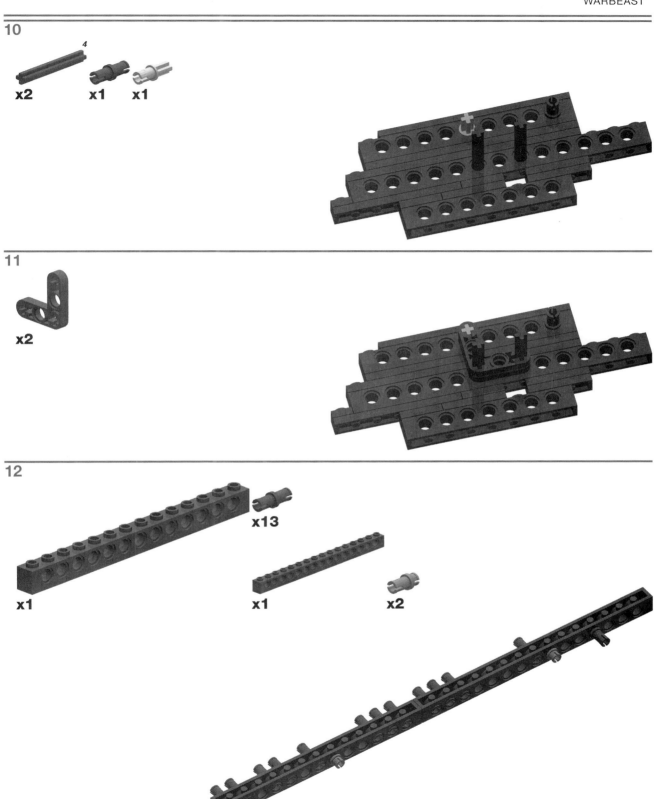

13

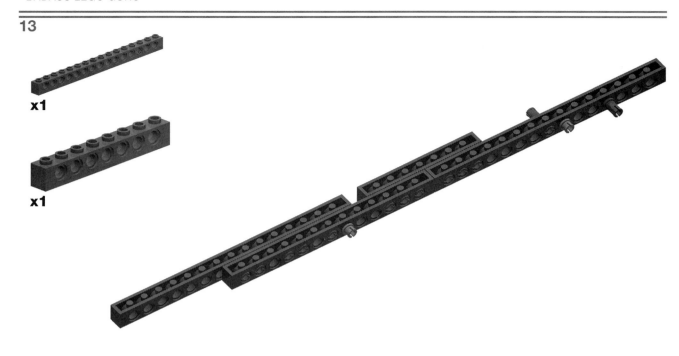

x1

x1

14

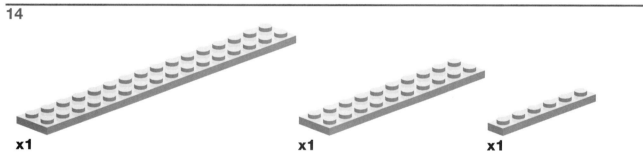

x1

x1

x1

15

x2 **x2** **x2** **x4**

16

x5 **x1**

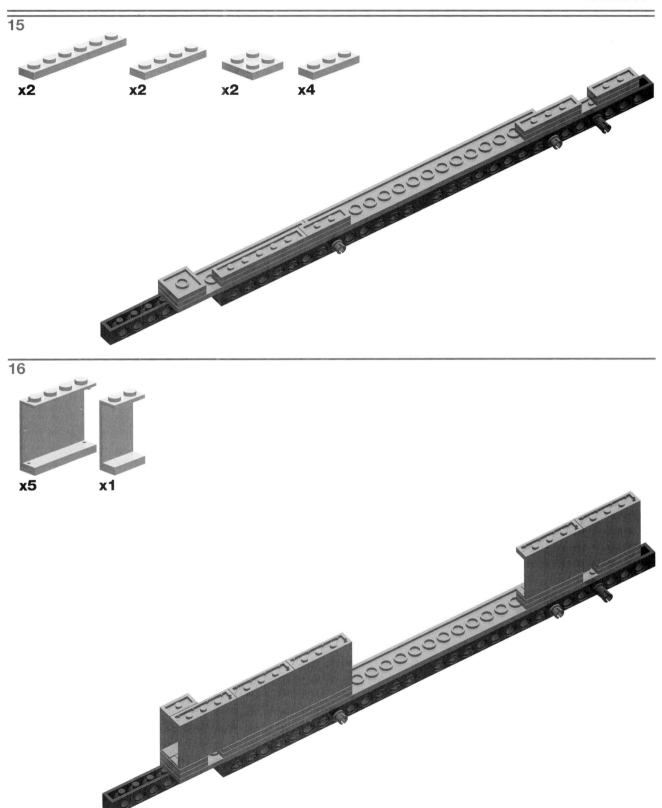

17

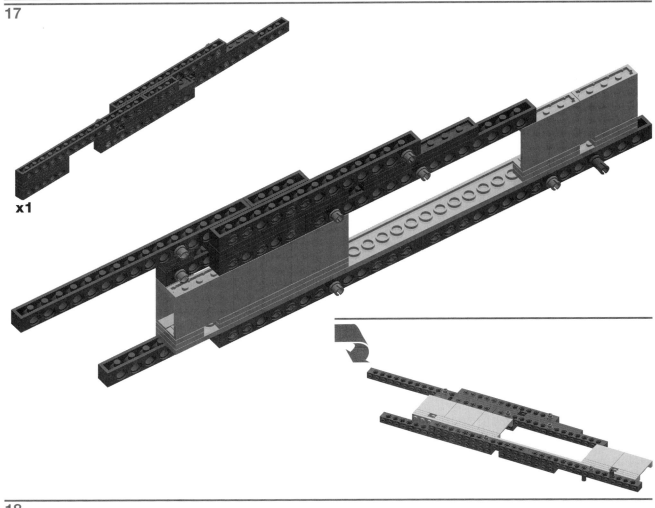

x1

18

x2

x1

x1

x1

x1

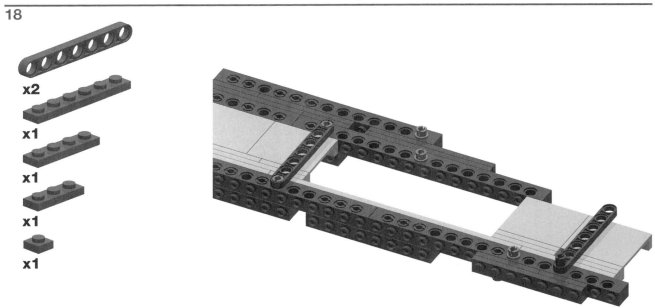

19

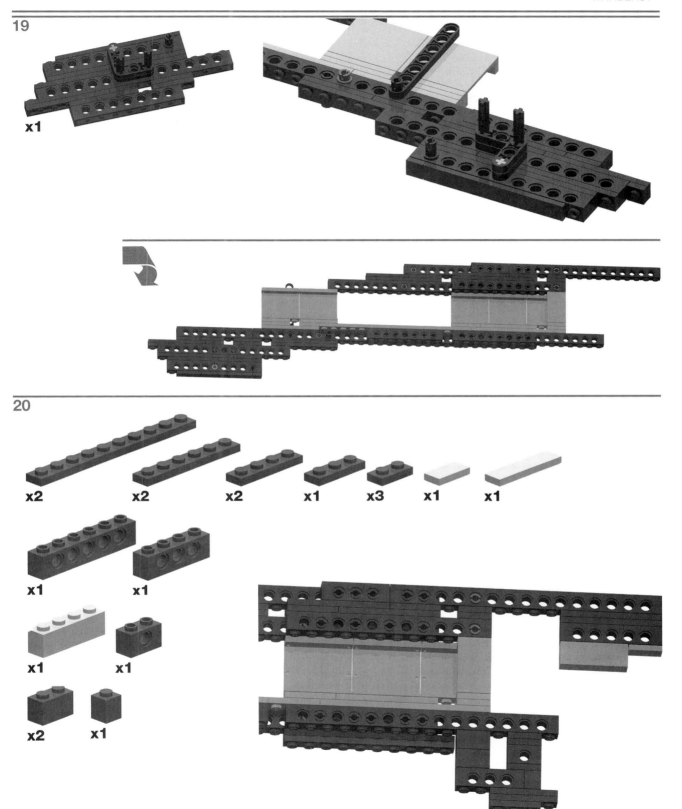

x1

20

x2   x2   x2   x1   x3   x1   x1

x1   x1

x1   x1

x2   x1

21

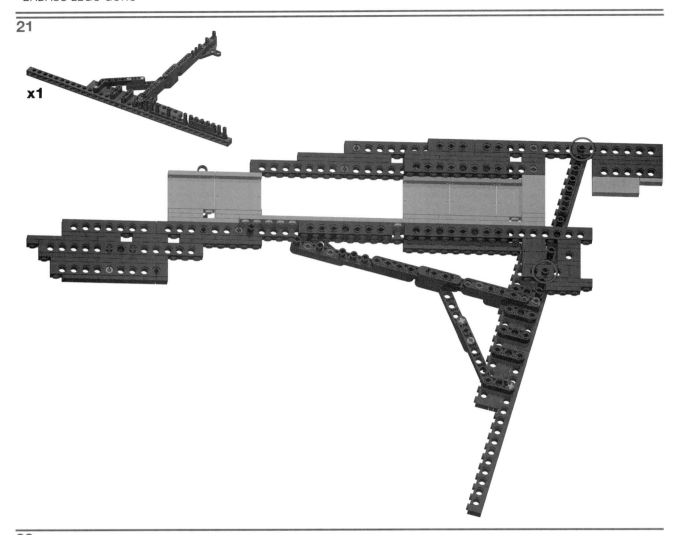

x1

22

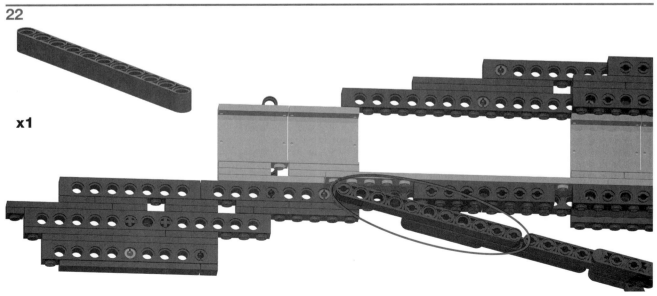

x1

23

x1 x2

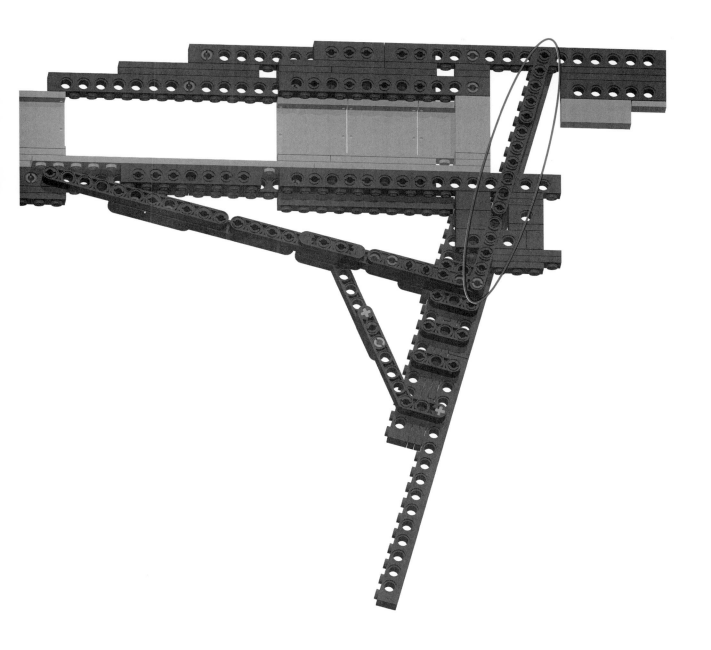

BODY: RIGHT SECTION

1

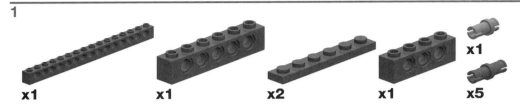

x1   x1   x2   x1   x1

x5

2

x1

x1

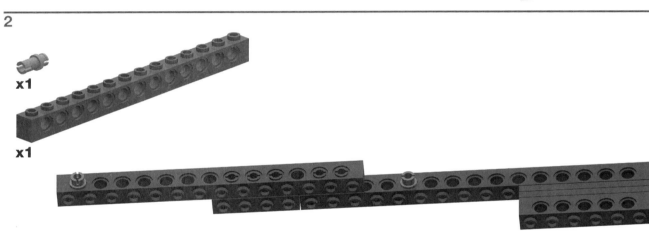

3

x1   x1   x2

x4

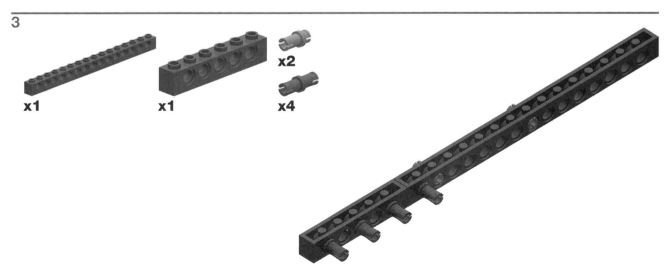

4

x1

x1

5

x2　　x2　　x2　　x2

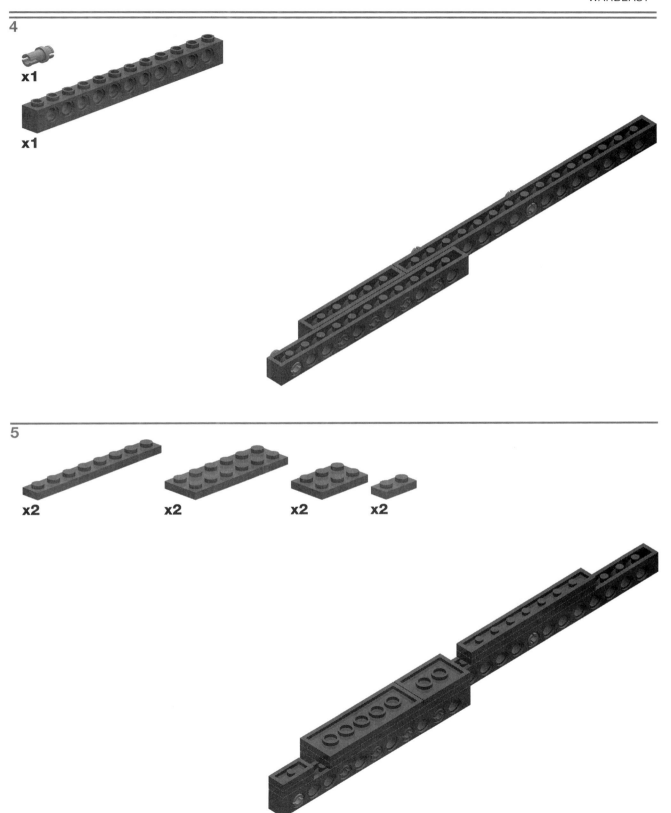

6

x1

7

x14

x1   **x1**

x1

8

x1   **x1**

x1   **x1**   **x1**

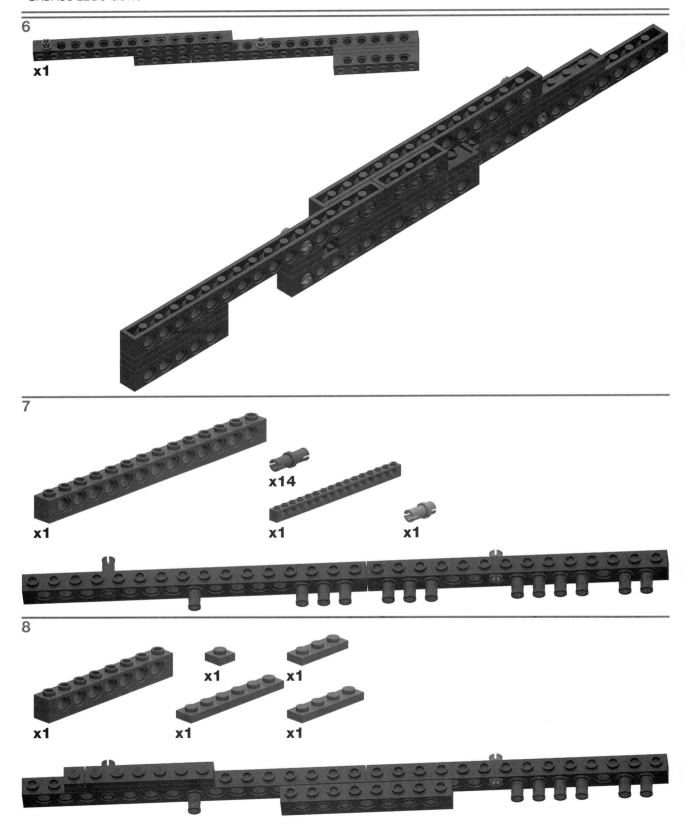

9

x1

x6

x1

x3

x2

x1

x1

x1

x1

x2

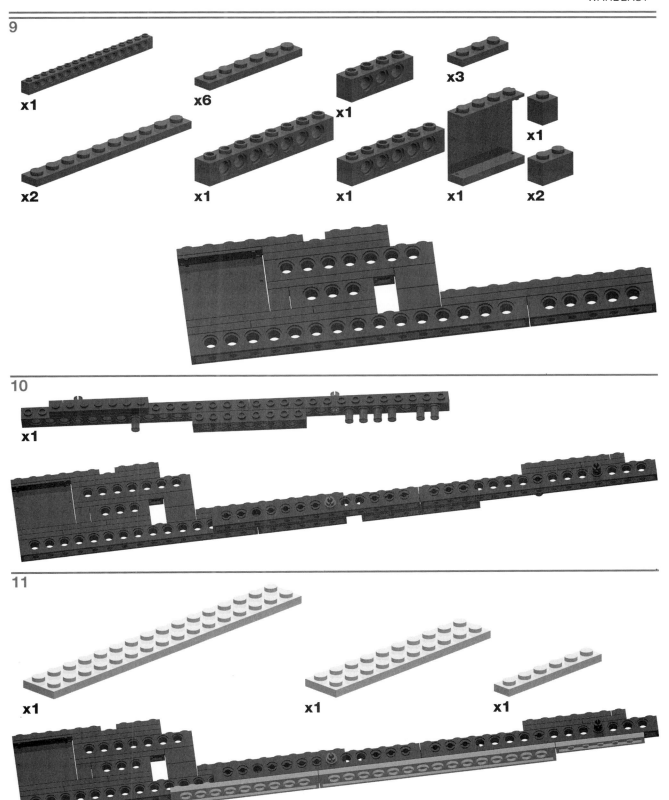

10

x1

11

x1

x1

x1

12

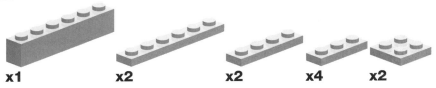

x1 **x2** **x2** **x4** **x2**

13

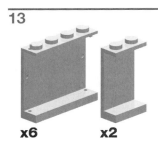

x6 **x2**

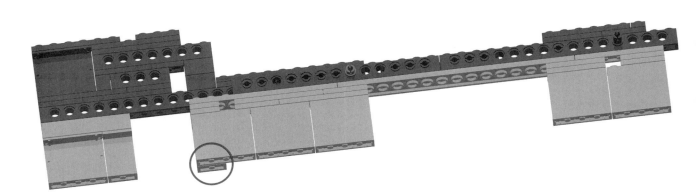

14

x1

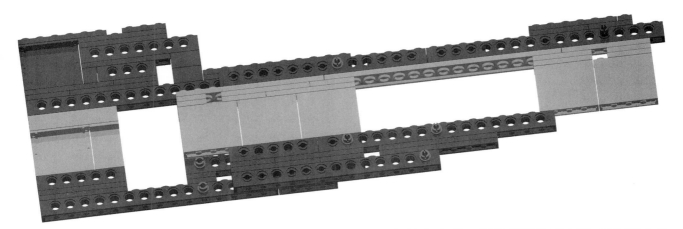

15

x2

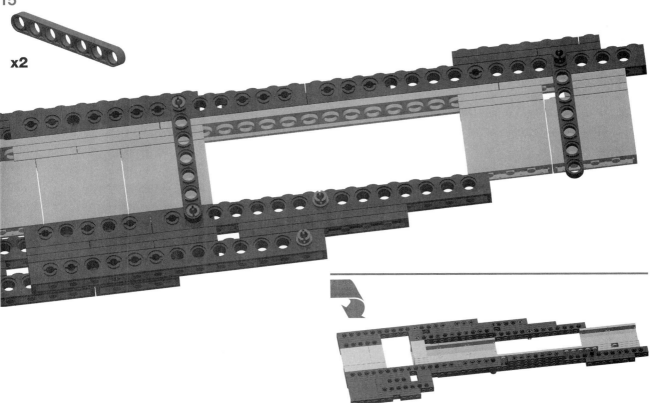

16

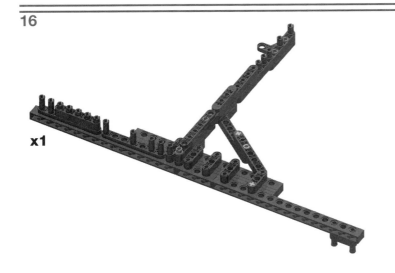

x1

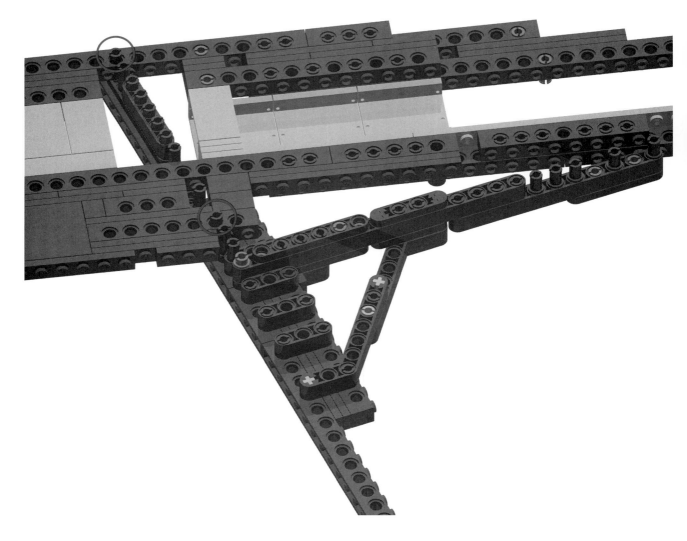

17

x1   **x2**

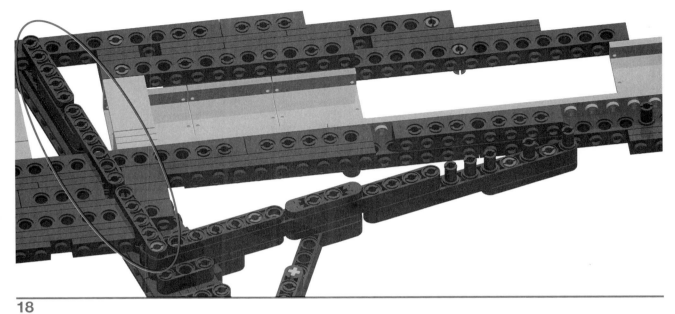

18

x1

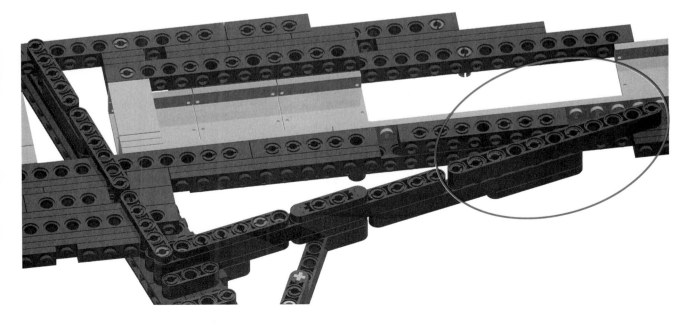

FRONT SHAFT

1

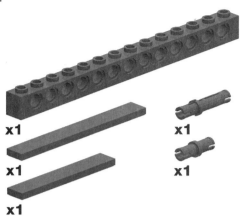

x1

x1

x1

x1

x1

2

x2

3

x2

x1

4

x1

x1

FRONT GRIP

1

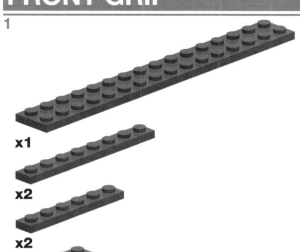

x1

x2

x2

x1

2

 x12

 x2

 x2

 x2

3

x1

FINAL ASSEMBLY

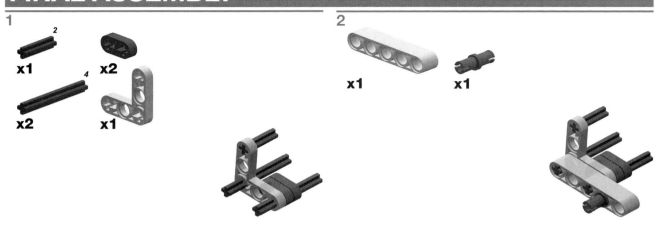

1

x1 2

x2 4

x2 x1

2

x1 x1

3

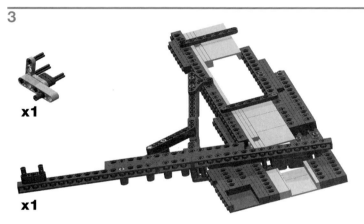

x1

x1

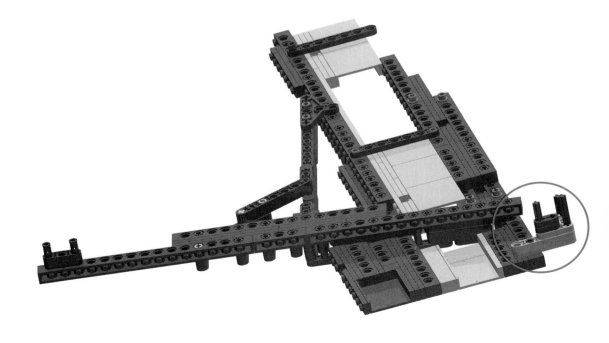

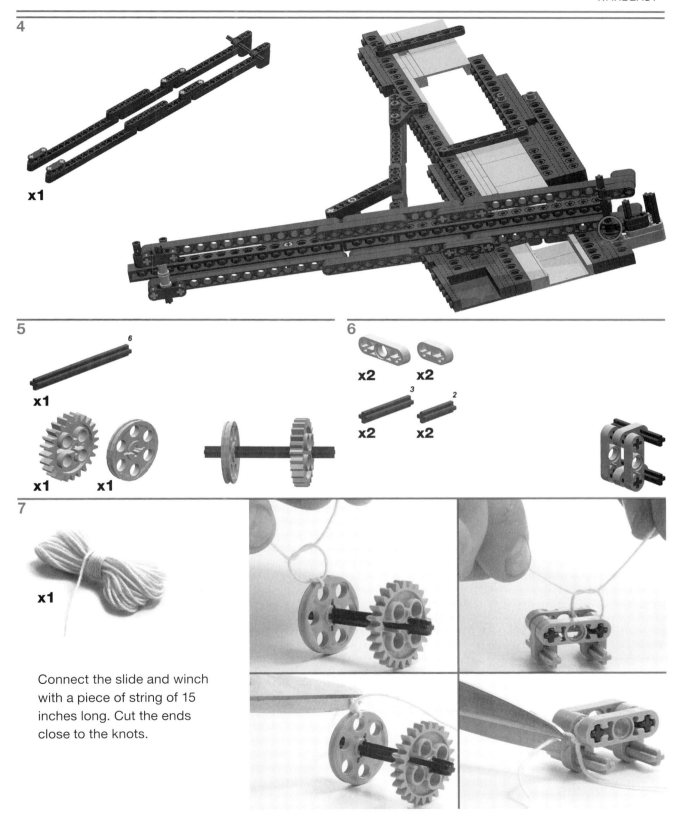

4

x1

5

x1

x1 x1

6

x2 x2

x2 x2

7

x1

Connect the slide and winch with a piece of string of 15 inches long. Cut the ends close to the knots.

8

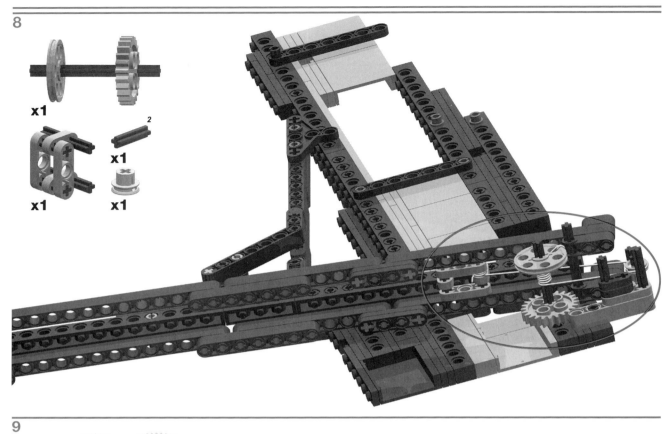

9

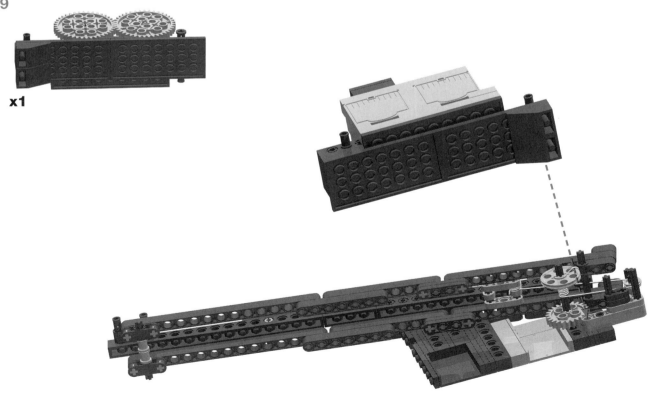

10

x1

11

x1

x1

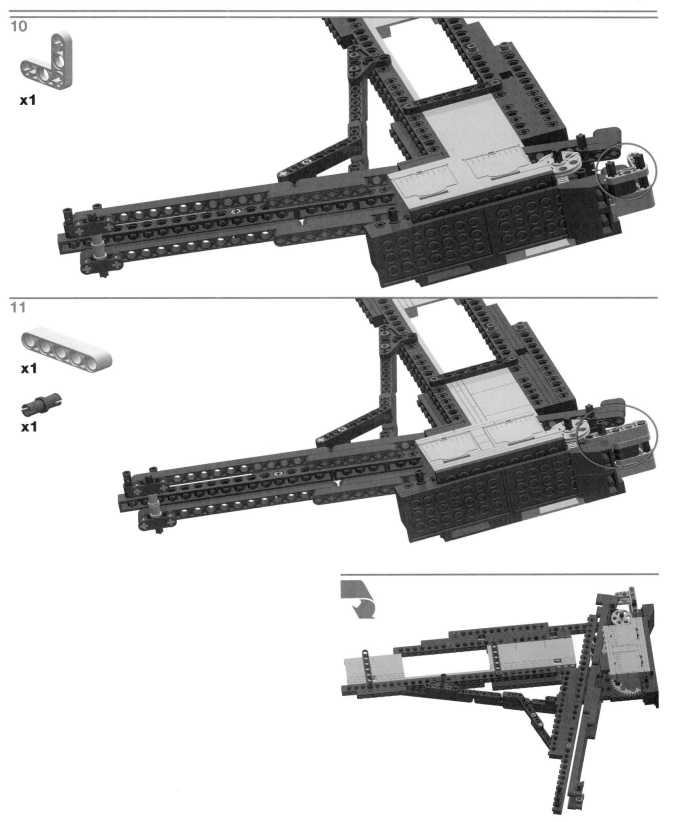

12

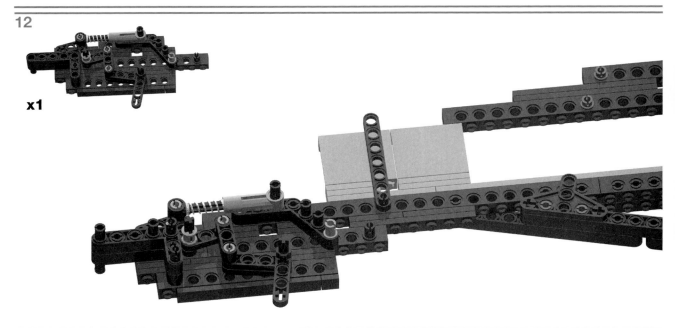

x1

13

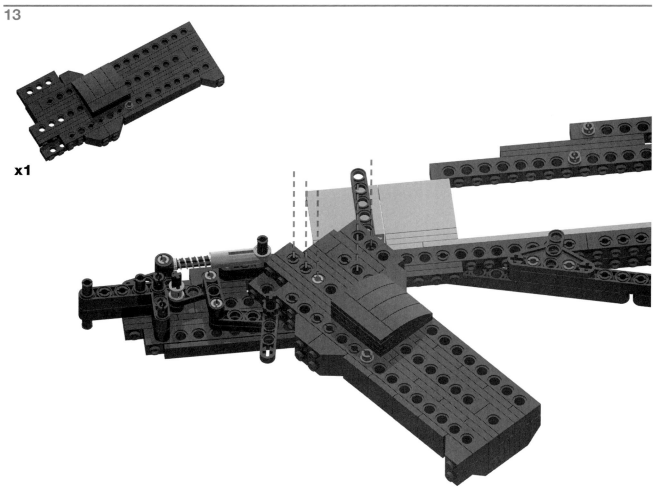

x1

14

x1

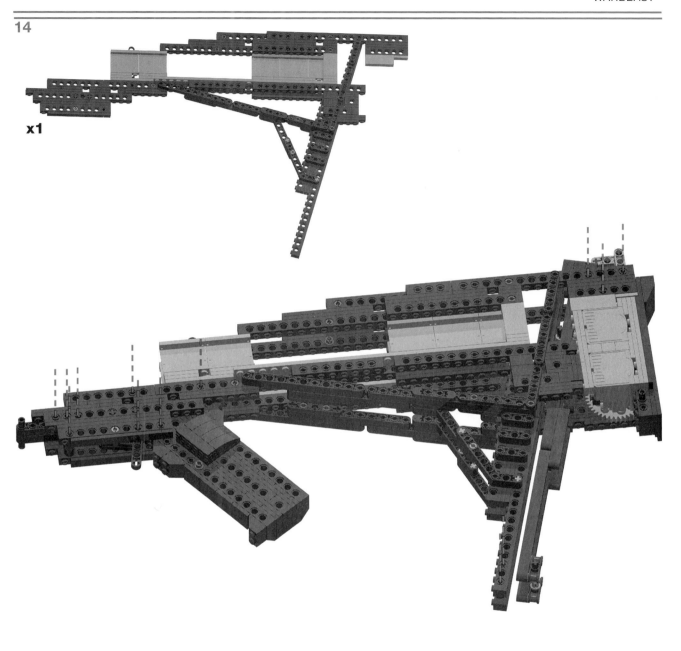

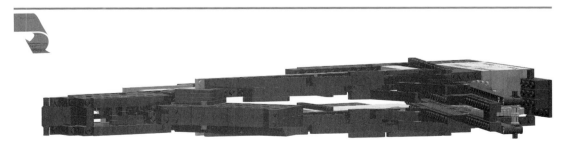

15

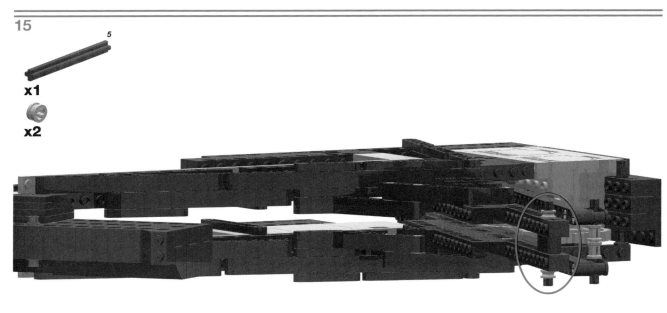

x1

x2

16

x2

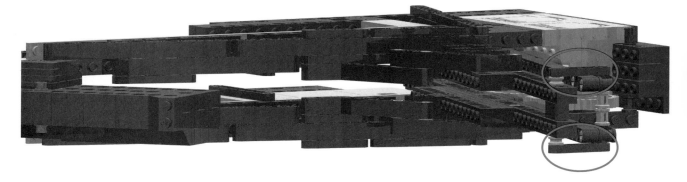

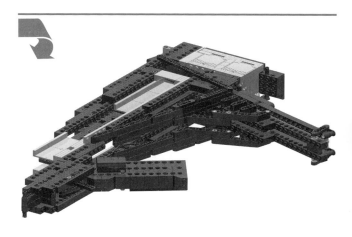

17

x1

x1 4

x1 3

x2

18

x1

19

x2 3

x2

x4

20

x1

x2

x2

21

x1

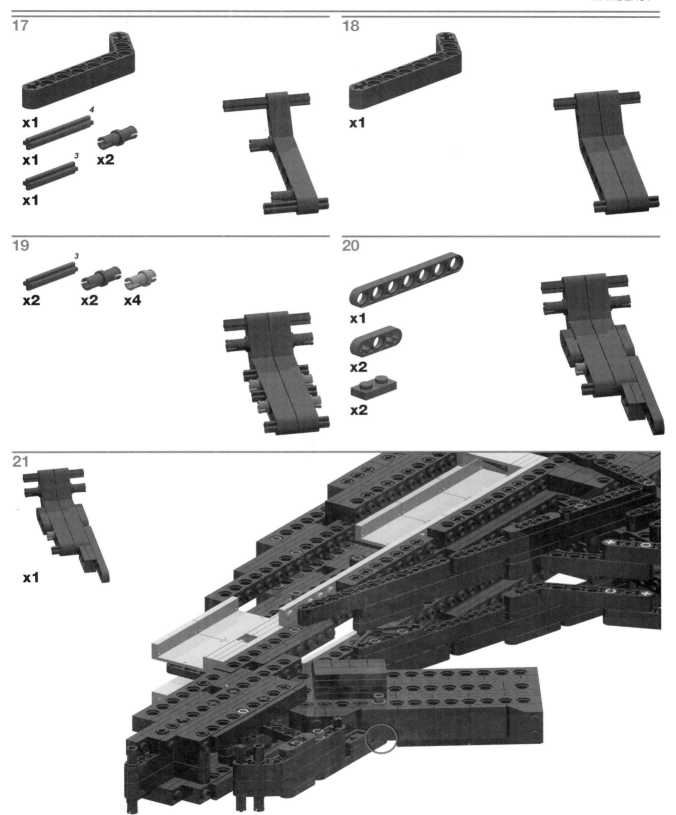

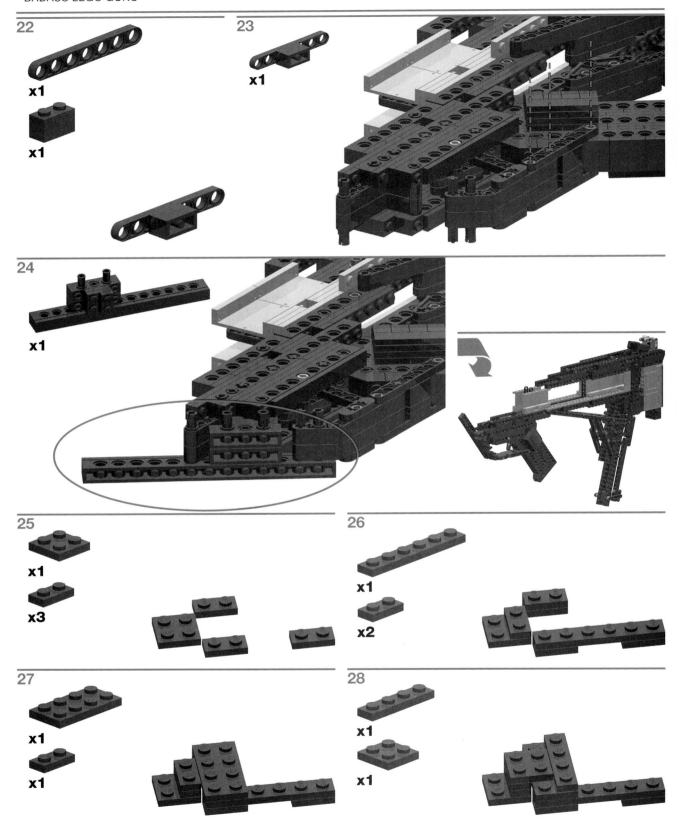

22

x1

x1

23

x1

24

x1

25

x1

x3

26

x1

x2

27

x1

x1

28

x1

x1

29

x1

30

69 studs length

x1

31

x1

32

x1

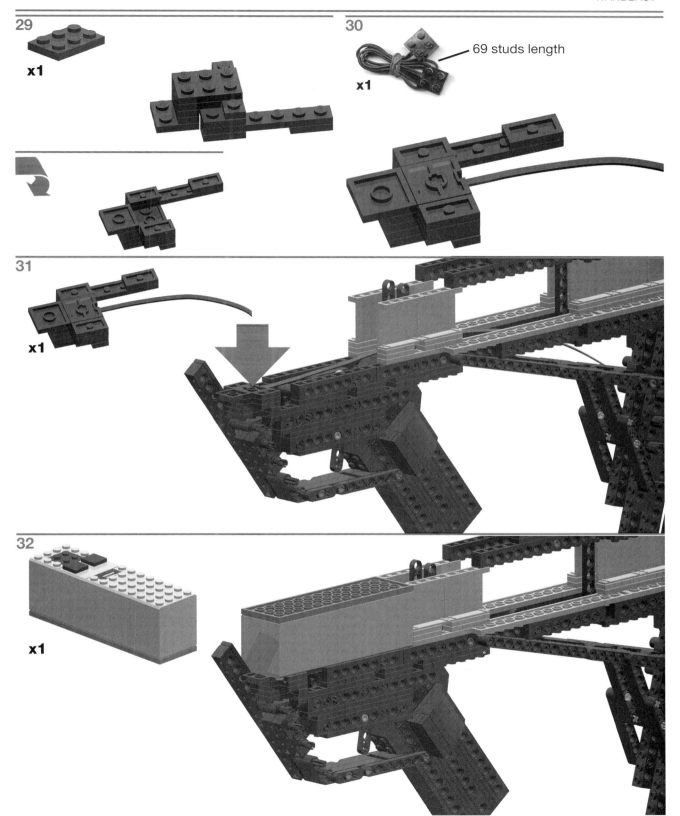

33

x1

34

x1 **x1** **x1**

35 4

x1

36

x1

37

x1

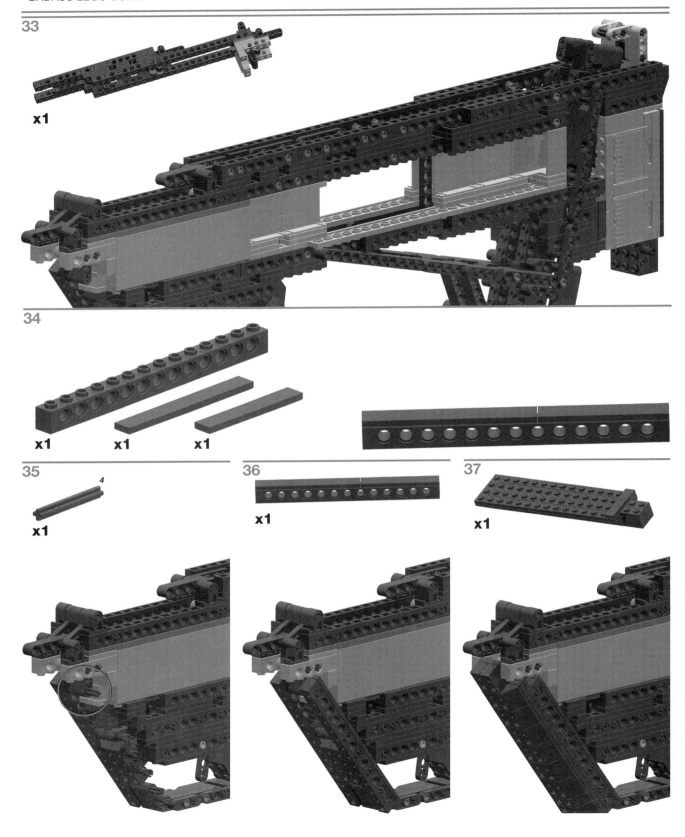

38

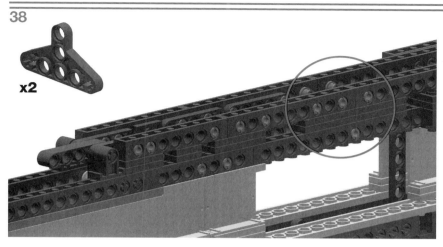

x2

In this step, you lock in the barrel that you attached in step 33 by using two Technic triangles.

Spreading the two sides of the gun with the help of a 2×1 LEGO brick will make it easier to mount the two triangles, as shown below.

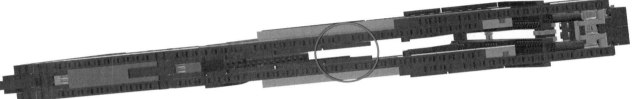

1

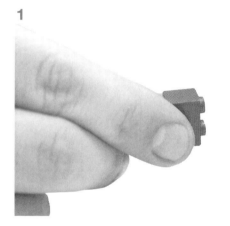

2

3

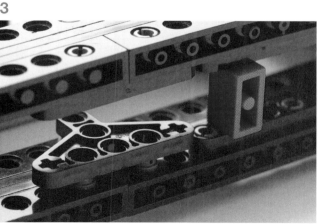

4

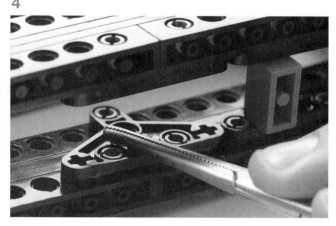

39

Now secure the two Technic liftarms (top, circled) to the barrel that you mounted in step 33.

Again, spread the two sides, and use tweezers to connect the liftarms.

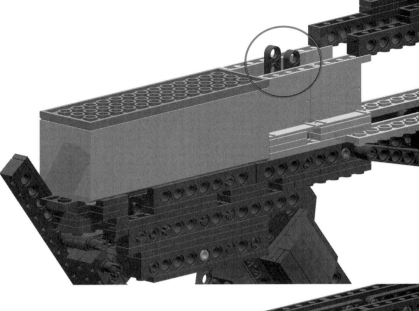

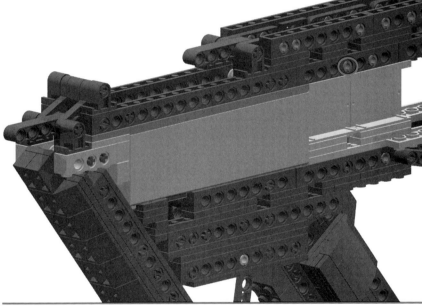

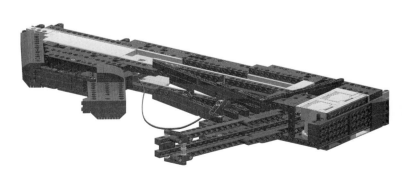

40

x1

41

x2

42

x5

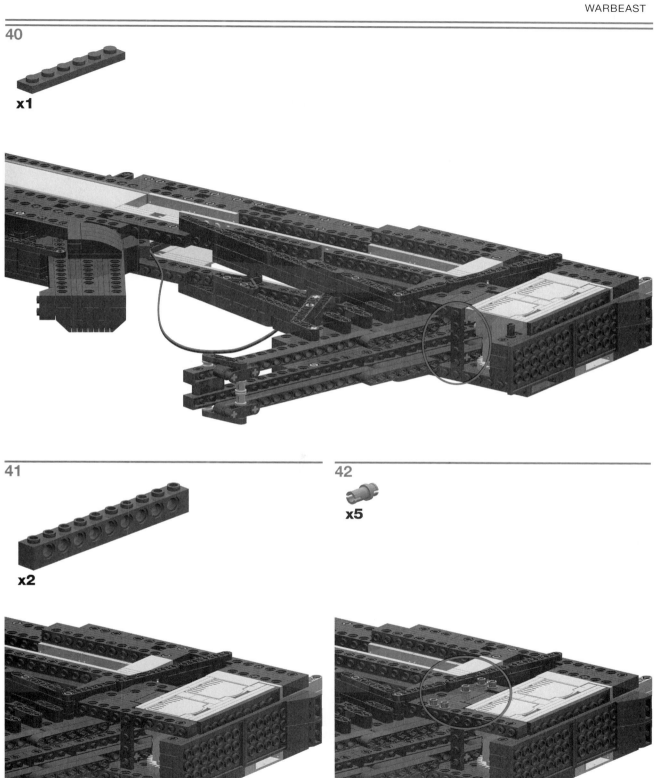

43

x2

44

x6

45

x2

46

2

x1

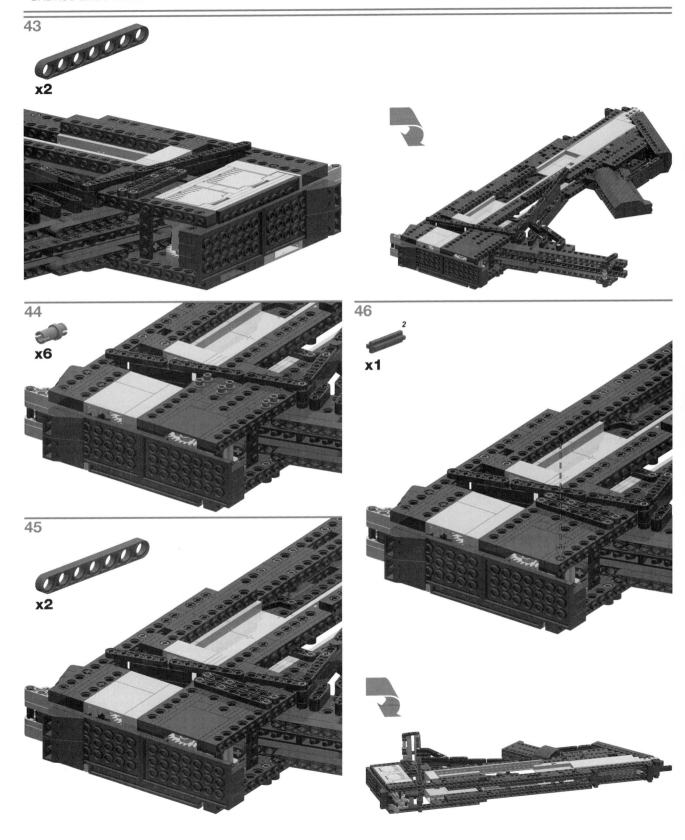

47

x2   x2

48

x1   x2   x4

49

x2

x2

50

x2   6

x1

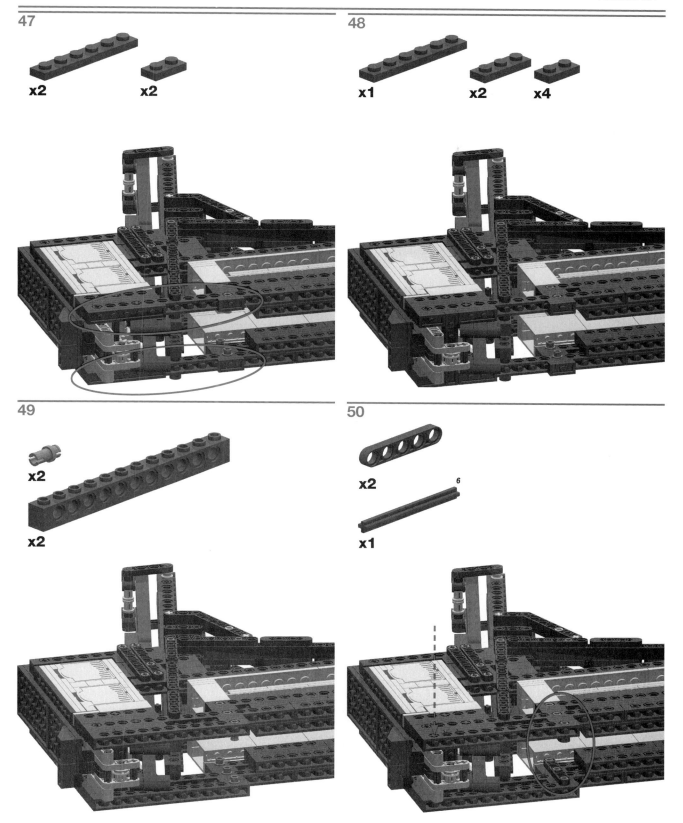

51

x1 x7 x1

x5 x1

x1

52

x1 x7 x1

x5 x1

x1

53

x1 x1

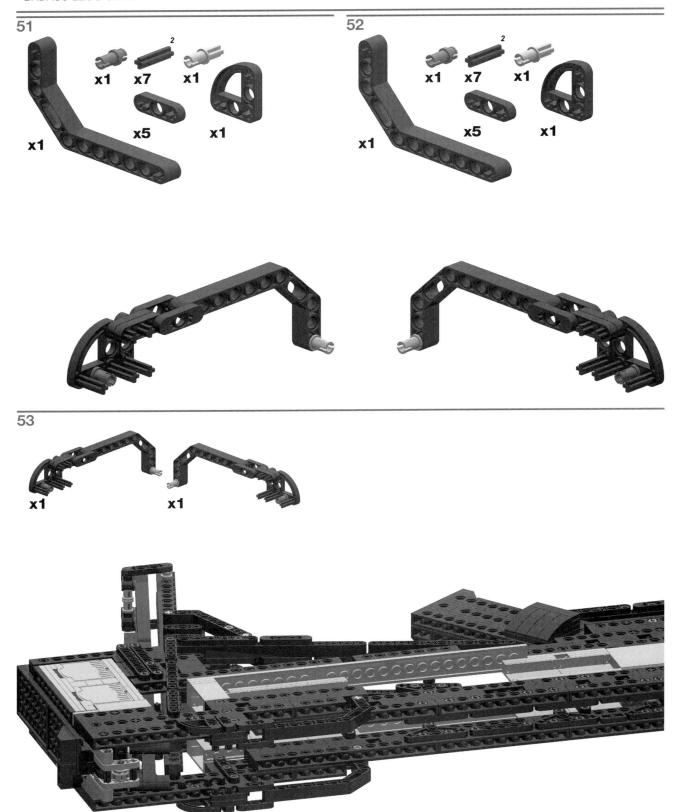

54

x2

x2

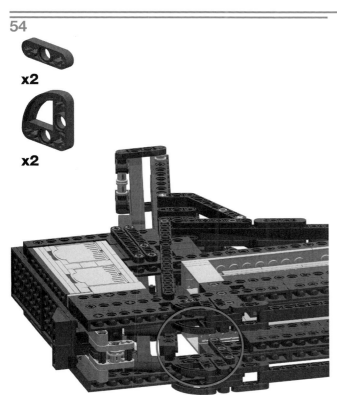

55

x1

Although this step is optional, oiling the mechanically moved parts will noticeably enhance the performance of the gun.

56

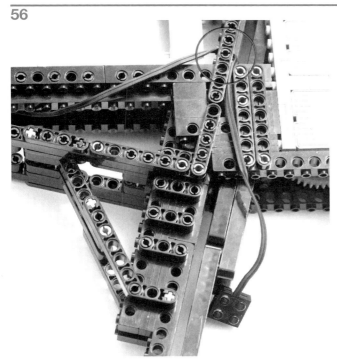

Installing the electric cable

LOADING THE GUN

1

85 mm diameter

x30

x30

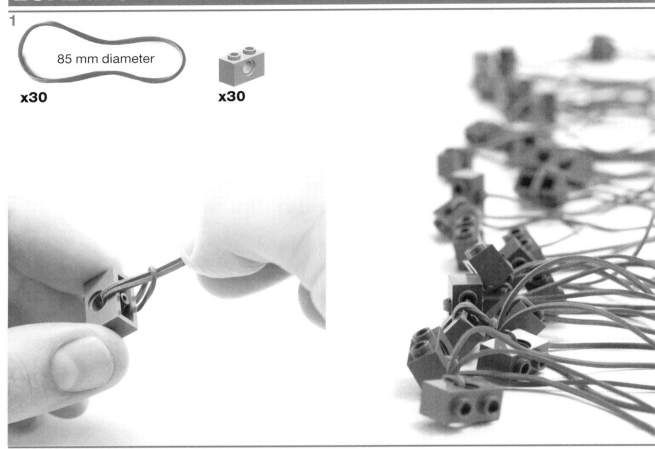

2

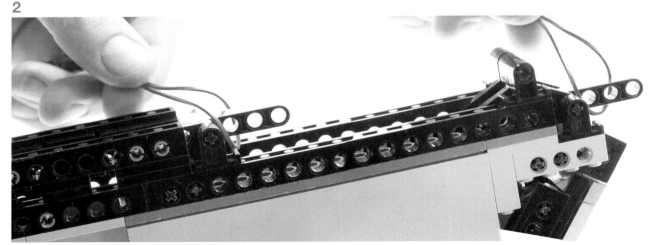

Loop the rubber band of the projectile over the front or rear rubber band catch. At first, newer rubber bands will fit only over the rear one.

But as time goes by, they will become more flexible and will fit over the front one, too.

3

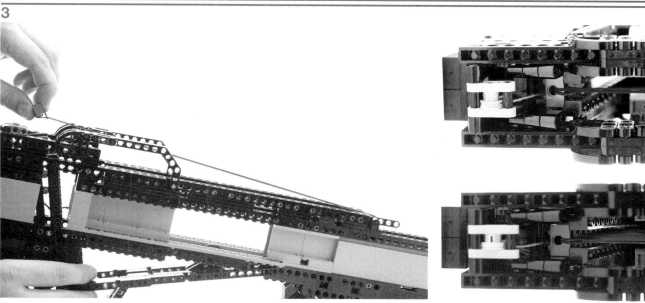

With one hand holding the gun, pull the 2×1 brick between the rails to the rear of the gun, and let it slide into the magazine.

Repeat this procedure until the magazine is fully loaded.

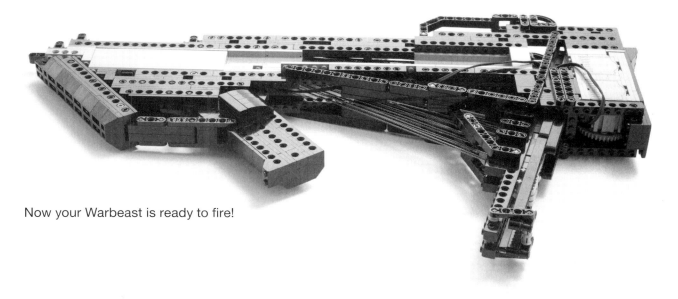

Now your Warbeast is ready to fire!

TUNING TIP: INSTALL A REFLEX SCOPE!

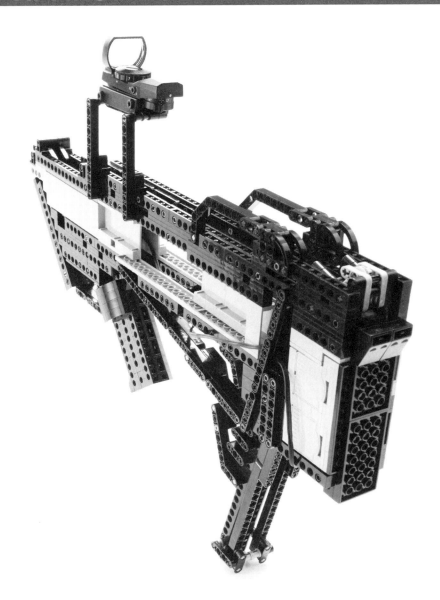

MAGIC MOTH

BUTTERFLY KNIFE

hinge

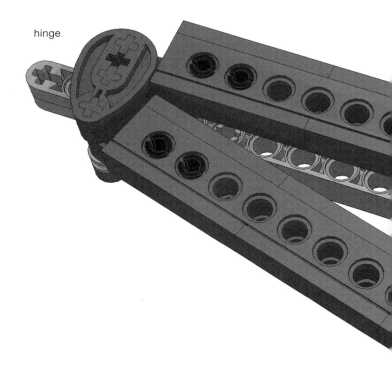

LENGTH: *15.44/25.76 cm*
WIDTH: *3.18 cm*
HEIGHT: *2.54 cm*
WEIGHT: *42g*
PARTS: *49*
LEVEL: *Beginner*

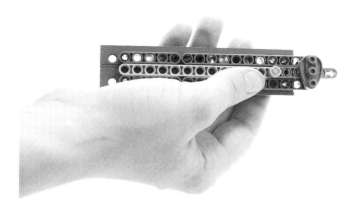

How to hold the Magic Moth

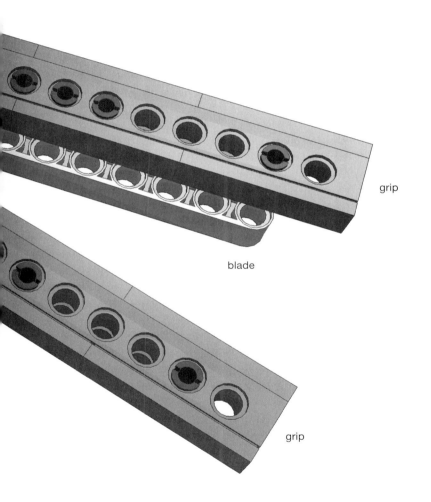

grip

blade

grip

The Magic Moth is a primitive butterfly knife. Of course, it doesn't have a real blade. Despite that obvious limitation, it's my favorite model: There is no greater fun than brandishing this gadget and learning tricks with it. Originally, the Magic Moth was top-secret. At the very last minute my publisher convinced me to release the construction plans and added them to the book.

BILL OF MATERIALS

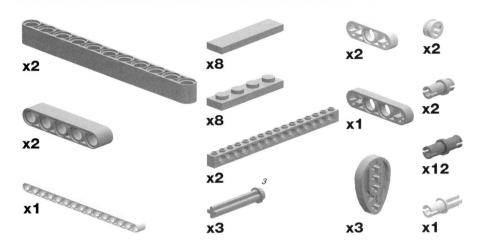

x2

x2

x1

x8

x8

x2

x3

x2

x1

x3

x2

x2

x12

x1

ASSEMBLY

1

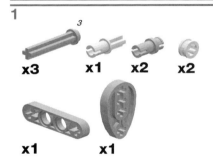

x3   x1   x2   x2

x1   x1

2

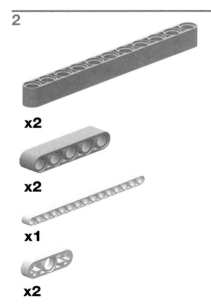

x2

x2

x1

x2

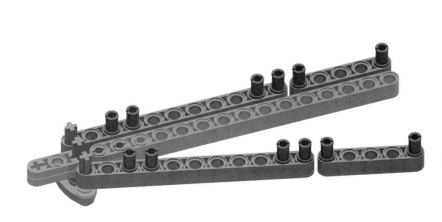

3

x12

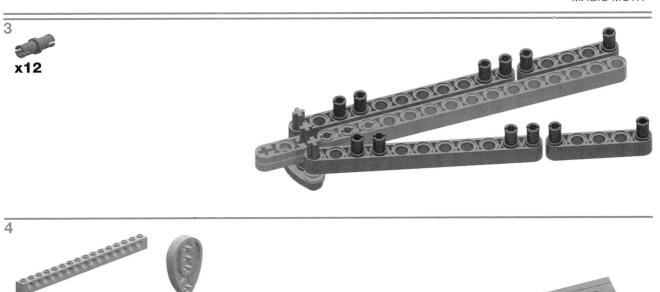

4

x2

x2

x8

x8

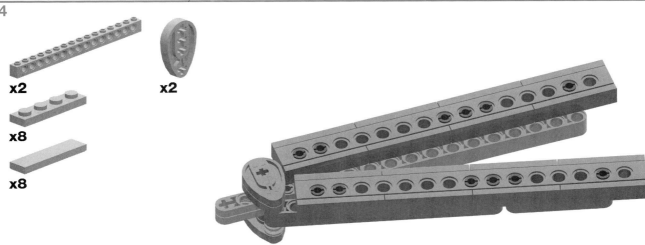

Badass LEGO Guns is set in Helvetica Neue. The book was printed and bound at Malloy Incorporated in Ann Arbor, Michigan. The paper is 80# Sterling Ultra Gloss. The book uses a RepKover binding, which allows it to lie flat when open.

The LEGO® Technic Idea Books
Simple Machines | Wheeled Wonders | Fantastic Contraptions
by YOSHIHITO ISOGAWA

The LEGO Technic Idea Books offer hundreds of working examples of simple yet fascinating Technic models that you can build based on their pictures alone. Colors distinguish each part, showing you how the models are assembled. Each photo illustrates a different principle, concept, or mechanism that will inspire your own original creations. The Technic models in *Simple Machines* demonstrate basic configurations of gears, shafts, pulleys, turntables, connectors, and the like, while the models in *Wheeled Wonders* spin or move things, drag race, haul heavy gear, bump off walls, wind up and go, and much more. *Fantastic Contraptions* includes working catapults, crawling spiders, and bipedal walkers, as well as gadgets powered by fans, propellers, springs, magnets, and vibration. These visual guides are the brainchild of master builder Yoshihito Isogawa of Tokyo, Japan.

OCTOBER 2010, 168 PP., 144 PP., AND 176 PP., *full color*, $19.95 EACH
ISBNS 978-1-59327-277-7, 978-1-59327-278-4, 978-1-59327-279-1

The LEGO® MINDSTORMS® NXT 2.0 Discovery Book
A Beginner's Guide to Building and Programming Robots
by LAURENS VALK

The crystal-clear instructions in *The LEGO MINDSTORMS NXT 2.0 Discovery Book* show you how to harness the capabilities of the NXT 2.0 set to build and program your own robots. Author and robotics instructor Laurens Valk walks you through the set, showing you how to use its various pieces and how to use the NXT software to program robots. Interactive tutorials make it easy for you to reach an advanced level of programming as you learn to build robots that move, monitor sensors, and use advanced programming techniques like data wires and variables. You'll build eight increasingly sophisticated robots like the Strider (a six-legged walking creature), the CCC (a climbing vehicle), and the Hybrid Brick Sorter (a robot that sorts by color and size). Numerous building and programming challenges throughout encourage you to think creatively and to apply what you've learned as you develop the skills essential to creating your own robots.

MAY 2010, 320 PP., $29.95
ISBN 978-1-59327-211-1

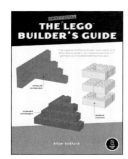

The Unofficial LEGO® Builder's Guide
by ALLAN BEDFORD

The Unofficial LEGO Builder's Guide combines techniques, principles, and reference information for building with LEGO bricks that go far beyond LEGO's official product instructions. You discover how to build everything from sturdy walls to a basic sphere, as well as projects including a mini space shuttle and a train station. The book also delves into advanced concepts such as scale and design. Includes essential terminology and the Brickopedia, a comprehensive guide to the different types of LEGO pieces.

SEPTEMBER 2005, 344 PP., $24.95
ISBN 978-1-59327-054-4